The Magnetic Double Helix, III

by

Herman Frederick Hagemier

FIELD THEORY

ISBN 0-75962-781-9

This book is printed on acid free paper.

1stBooks - rev. 10/11/01

PROLOGUE

John Gribbin in his 1997 biography of Richard Feynman, "A Life in Science," tells us that, in many ways, Feynman's electro-dynamics became classical. By using his diagrams Feynman multiplied Einstein's space time coordinates in such a way that they became three dimensional instead of four. Is that what Gribbin meant when he said that the theory became classical? In chapter 12, page 55, of my book, "The Magnetic Double Helix," I make a specific suggestion on how this return to the classical could be accomplished. It is quite possible that Feynman used a different multiplicand but that would be one way to return to the classical.

Gribbin said that Feynman and John Wheeler, his thesis advisor, realised that what they had, from then on, was classical mechanics. They marveled that it was so simple to work with. It had a directness that solved many problems.

On pages 88 to 89, Gribbin highly praises Feynman's electro-dynamics and classical mechanics and scathingly excoriates the world's physic's establishments for what they do to their students. Almost all are forced to learn physics the old fashioned way, and then they have to learn Schrodinger's wave approach along with the Haniltonian mathematics. If a student has already learned two ways to do a job, learning a third but easier way will not give him back the time he wasted on the first two. This is too much like a union restricting the number of qualified workers. This state of affairs may result in Feynman's avoidance of that worst kind of nonsense being gradually forgotten.

Feynman has told the story a number of times that, when he started working on quantum theory, he wanted to be ignorant of, or at least disregard, what others had done before him. Reading Feynman's statement reminded me of what the brash young physicists had done in the early twentieth century when they decided to ignore the rules of common sense and disregard most of the history of the nineteenth century.

Thomas S. Kuhns, in his book, "The Structure of Scientific Revolutions," tells how he had completed his studies in physics to the point that he was ready to write a thesis, when he accidentally discovered what the physicists had done to their history. Many kinds of revolutionary problems had been swept under a rug. The physicists suppressed almost every kind of history that might put questions into class discussions. Kuhn was indignant over such student treatment and changed his studies from physics to the history of science. Thomas Kuhn's 1962 book is still popular, but probably not popular with the physicists.

Feynman was not overly conceited when he decided to ignore the then present state of quantum theory and take a fresh start. It was because he had an extraordinary ability to see new ways to solve problems. He had often been admired for those skills. He had a keen appreciation for common sense and the quantum argument that, "This is an uncommon sense universe," would have had little appeal to him. Getting rid of the uncertainty theory and the acausal theory was a way that he could simplify quantum theory tremendously and there is a lot of evidence that that happened. By putting time into his equations he got rid of Hamiltonian mathematics and going to classical mechanics was another big step toward simplification. Gribbin seemed to appreciate Feynman's stand on simplicity.

In the biography by James Gleick, "Genius," page 8, Feynman is described being criticized and humiliated by Neils Bohr at a group meeting in 1948, because he had left the uncertainty out of his quantum theory. We can understand why men like Bohr, who helped create the theories, would jealously guard them. It is difficult to forgive them, however, for still suppressing Feynman's much simpler mechanics, fifty years after Feynman created it. In this prologue for his book, James Gleick says that the conservative quantum dynamics that these physicists were defending gave answers so wrong that they were senseless.

If this author is reading Gribbin right, these are the theories that they are still teaching in their schools. They are still teaching that this is not a common sense universe. This must be helping the spread of every kind of paranormal nonsense. Has physics become paranormal also?

Feynman resented the way-out groups who used uncertainty theory, Einstein's theory, and bits and pieces of quantum mechanics to adorn science with their own presence. He called it "hokey-pokey," or fairy tales and wanted to debunk that kind of stuff. He voiced these thoughts in talking to a friend, Faustin Bray, at Esalen in California. See on page 97 of "The Illustrated Richard Feynman. No Ordinary Genius," edited by Christopher Sykes.

As the author of this book I have no reason to be conceited about my feat of proving both Lorentz and Einstein wrong in their assessment of the results of the Michelson-Morley experiment. The algebra was very easy and I was using the proper theory of an ether. That was the multidirectional and energetic ether which was invented by George LeSage is 1744. That was also the ether theory which Clerk Maxwell almost killed by saying that the energy it would contain would burn up the universe. This author would endow the ether with magnetic energy. That should meet Maxwell's objections. Sir J.J. Thomson was very fond of that ether. He used it in 1907 in his theory for the creation of energy fields.

Sir Edmund Whittaker in Volume I, page 366, of his "History of the Aether and Electricity," claimed that Sir J.J. Thomson was the foremost experimental physicist of his time. Graduates of other universities came to Cambridge to work is his laboratory. His reputation and their work helped his students obtain important chairs in physics all over the British Empire. He was famous for proving the existence of an electron by weighing one..

Whittaker, Vol. 11, pp. 247-9. The continual dispersal by electrons of bodies moving with velocity C in all directions was part of a theory of electromagnetism put forward by J.J. Thomson in 1907. This is a way in which a LeSage type ether can build fields of force.

J.J. Thomson in, "Recent Researches in Electricity and Magnetism," Clarendon Press, Oxford, 1893, page 15, tells us that the momentum used by the Faraday tubes would be furnished an ether similar to the LeSage's gravitational ether. He included the oppositely moving waves. This author has been saying that an ether which comes from all directions and leaves in all directions, can flow in all opposite directions at the same time.

Ibid., p.29 Thomson supposed that there are as many negative tubes as positive tubes for each unit area of the field. The positive tubes are moving at the speed of light in one direction and the negative tubes are moving at the speed of light in the opposite direction.

Ibid., p.35, J.J. Thomson had the concept of a fan-like mechanism which would drive the positive and negative tubes in their opposite directions within a bar magnet. Is there much doubt that, had he known of the electrons in magnetic orbits, at that time, he would have ascribed to them that fan-like action?

"The Magnetic Double Helix, III" is by H.F. Hagemier. The scale drawings on the last three pages help explain the contractions that Einstein couldn't explain. There are two lines of force sent out by every magnetic orbit. They leave in the two directions perpendicular to the planes of those orbits. The orbits are more or less circular. The electrons send out helical waves which travel in the two directions as if the opposite waves were not there. The length of the magnetic waves is such that a helix from one direction, a helix from the opposite direction, and an electron in its orbit, meet in three way intersections. Ether enters the north end of a magnet and a magnetic wave exits the south end. Ether enters the south end of a magnet and a magnetic fie1d exits the north end.

All bonds exercise their pull when they reach their destination. It is the same with gravitational bonds. There is no such thing as a push and an electron cannot act upon itself. It can respond to vectors of force from other directions.

The ether which can carry light in all opposite directions can carry the two-way bonds also. Momentum is created by the inverse square effect. See chapter 2. A wave to the front is longer and weaker. A wave to the rear is shorter and stronger. In the drawings, an overlap of one space from the rear shortens and strengthens the bond from the front. If x is the number of wave lengths from the proton, x^2 is the number of spaces in the wave lengths in the drawings.

There is the inverse square pull which drops with the number of wave lengths from the hydrogen proton, squared. We divide by, 1, or 4, or 9, or 16, or 25, or 36, etc. If the contraction of the light path in the line of motion is $1 - V^2 / C^2$, the contraction in lateral direction must be $(1 - V^2 / C^2)^{1/2}$. The 1932 Kennedy-

Thorndyke experiment rules out the ratio of the two contractions which Einstein used.

The counting of the spaces will always be straight across the page. The contraction for the wave will be one space, from Bt6 to Bt5. $V^2 / C^2 = 1 / 16$. The contraction for 1 / 1000 of the speed of light would be one millionth of a wave length.

The Michelson-Morley experiment was comparing the round trip speed of light in one direction with the round trip speed of light in a 90° different direction. The experimental aparatus was floating and could be rotated to point in a full circle of directions. One direction will be the one in which the earth and the laboratory are moving. Hendric Lorentz changed the mathematics of the M. M. experiment by suggesting that instead of using the two contractions, we should use only one contraction, which will be the ratio of the two contractions. The two light paths were the same length and the contraction in the direction of motion was the square of the contraction in the lateral direction. The ratio of A2 / A = A. Lorentz used the ratio as a contraction in the line of motion and also offered the same contraction on a hypothetical time axis. and did away with the contraction in the lateral direction.

Einstein used the changes that Lorentz had made in the M-M experiment format in constructing his relativity theory. That makes the Kennedy-Thorndyke experiment of 1932 important. That experiment was performed using the Michelson-Morley laboratory interferometer. The big difference was that two light paths were widely different in length. The K.T. experimental results were null. That was the same result that M-M received. The ratio of A^2 / A will be A. The ratio of A^2 / B can not possibly be A unless B is equal to A. In this case we knew that there was an intentional difference between the lengths of A and B.

PREFACE

A 1992 book, "The Magnetic Double Helix," is being enlarged by adding several chapters. In the first chapter there is a preliminary discussion of the field equation for the momentum of the hydrogen electron orbits and the momentum of the planetary orbits. This is truly a unified field equation. We can now compare the gravitational pull of the sun on a planet to the pull of a proton on an electron. The field prepares the sequence of the orbits for electrons and planets. This is just one breakthrough among a score of others. During the time of the formation of the planets there were so many collisions that more planets were destroyed than there are planets which survived.

The length of the radius of the orbits of Neptune and Pluto were badly skewed in the Bode's Law that we have been using since the above planets were discovered. The new Bode's Law has corrected that problem. See chapter five on the planetary orbits.

The first chapter deals with the field equation. The second chapter discusses the discovery of straight-line momentum. The third chapter is about the Balmer Formula. The fourth chapter discusses the quantum orbits. The fifth chapter covers several. aspects of the planetary orbits. The sixth chapter, with several additions, is the introduction to the 1992 book, "The Magnetic Double Helix." After the sixth chapter the rest of the 1902 book will follow with the 1992 sequence of the remaining chapters.

ACKNOWLEDGMENTS

I am very much in debt to Prof. Howard Hayden who warned me that I was wrong by a factor of 4 in a series of values in a paper I had submitted to the bimonthly, GALILEAN ELECTRODYNAMICS, in Storrs, Conn., in February 1995. That was the first time in thirty-four years of constantly reading in the history of physics and relaying my discoveries, that I have had encouragement from a physicist. I hope that the code I used was not a factor.

To D. S. Parasnis[34] I owe much. He kindly gave me permission to quote from his book, "From Lodestone to Polar Wandering," He had been puzzled about a number of facts about the distribution of magnetic strength along a bar magnet. Parasnis published significant facts that others completely ignored. He explained that slots could be cut into a bar magnet to measure the field strength at many points. The magnetic double helix explains magnetism like it had never been explained before.

I thank my wife, Lou, for her patience. I thank Dr. Keith Ault who gave technical aid and encouragement. I thank Mark Hughes who gave computer expertise and other support. I thank David Helton who helped with the drawings. I thank Phillip Blumenthal who found books for me. I thank the used book stores where I found rare history. I thank the librarians who found scientific papers on faraway shelves. I thank the historian, Thomas Kuhn[5], for, "The Structure of Scientific Revolutions."

Pendleton, Indiana Herman F. Hagemier

1999

Table of Contents

Chapter 1

The Revolution and the Unified Field Equation

This author is not a Don Quixote, tilting at windmills. This is for real. This book describes a common-sense universe which will replace the uncommon-sense universe which relativity theory and quantum theories describe. Many millions of persons can understand the simple algebra that is used in describing this common-sense universe, but only when this common sense is made available to the millions.

As far as the physical sciences are concerned and as far as the defense of logic is concerned this may be the most important book published in this century. The assault on common sense and logic by the frankly unashamed attacks by relativity and quantum theories has eroded support for all kinds of critical thinking. Science sets a bad example and the mystics are flourishing everywhere as a result.

When the Michelson-Morley experiment is explained by a two-contraction theory, the foundation disappears from Einstein's relativity theory. There is an explanation for the contractions of mass in motion. The hydrogen orbits and the planetary orbits can be explained now by a solution for that greatest of all mysteries, momentum at less than the speed of light. Before this chapter is over you will be introduced to an equation for a unified field theory. This is a field that is shared in its simplest form by the hydrogen electrons and the sun's planets. This is not the unified field that Einstein was seeking. He was imagining a unified field that would include an extra dimension of space and time and a speed of light which could never be relative to another speed. Late in his life Einstein was beginning to see the need for an ether. This author sees the ether as tiny magnetic particles, moving at the speed of light, and so small that they could travel mean free paths across the known universe.

Magnetism can be explained by double-helical magnetic bonds. There is an addition to gravitational theory that puts time into Newton's laws. See what that does for Mercury's perihelion. There is a new Bode's law for the planetary orbits, the first real change in three hundred years. The inverse square law is placed solidly in the field force for the new Bode's Law and the quantum orbits. The physicists developed theories about quantum orbits which are based on already proven physical facts. They had been collecting facts about the spectra of the elements for over two hundred years. It is no wonder that the facts fit the mathematics of their theories, the mathematics were chosen to fit their interpretations of those facts and/or the interpretations were chosen to fit the mathematics they had invented.

Glaring inconsistencies in their theories are freely admitted by the physicists. They can explain them all by the statement that this is not a common-sense universe.[45] If you insist on a common-sense universe, you are labeled as being naive and ignorant of the fine points of modern science. They will point out how well the mathematics fits their theories.

P.S. The following are examples of uncommon sense.

A most serious example of a lack of common sense is the manner in which scientists are wasting thousands of years of research time chasing a willow-o'-the-wisp called, "The Big Bang Theory." Hundreds of books and papers are being written each year. This is according to Eric Lerner[46], in his book, "The Big Bang Never Happened," which was published by Vintage Books in 1992. We can speculate that the Big Bang Theory is an outgrowth of a strong desire to return to the finite universe of creationism and to turn away from the infinite universe which scientist have favored since Aristotle's philosophy was deposed about four hundred years ago.

Eric Lerner reasoned that the Big Bang Theory would continue to be tolerated by the physicists as long as nothing else was offered. He found that he had a theory, already shared by others, that a large part of the universe is composed of vast streams of plasma, (electrical particles.) Cosmological observations and tests have confirmed their ideas in many ways.

Eric Lerner tells of large groups of galaxies showing evidence of having been in existence billions of years before the Big Bang was supposed to have taken place. Also there is a shortage of the mass that to create the gravitation that would be necessary for the Big Bang to have formed the present universe. The Big Bang Theory needs a hundred times as much invisible matter or dark matter as there is real matter. There is none.

The Big-Bang physicists have faith that this little problem will be solved sometime. So far, all attempts have failed, despite the newspaper headlines which appear at irregular intervals to say that at last the missing mass has been found. Meanwhile they have joined with the particle theorists in urging the construction of ten billion dollar atom smashers. Just maybe they might discover something to help them in their search.

Another author who mentioned the cost of atom smashers was Ruggero Santilli[47] in his book, "Il Grande Grido." Working at Harvard as a particle physicist, he was terminated because he proposed an experiment which might reveal that the mathematics of relativity theory was unsuitable for use in particle physics. The mathematics of relativity treats particles as being mathematical points. A point particle divided into three quarks may have presented Santilli with some problems. The big problem, however, seems to be the protective censorship.

We can return to Eric Lerner. He mentioned an interesting problem in the physics of special relativity where it says that an electron is a point particle.

Because an electron is infinitely small and because energy increases as the distances become smaller, the electron has infinite energy and weight. Richard Feynman[48] and two others received the Nobel Prize because they gave the electron its true weight in their calculations. Feynman called renormalization a dippy process, a sort of hocus-pokus, and violated their own rules, but they got rid of the infinities and a theory became useful.

Nothing in this book is meant to imply that all physicists are satisfied with the present state of our science. Many have said that nothing less than a complete revolution will straighten out the basic problems, but they see little hope. As Lerner suggested, they do not want to get rid of what they have until they have something better. To disprove it and not provide something better would embarrass the teaching of science.

The Field Equation

Sir Edmund Whittaker[18], in his history, told of Sir J. J. Thomson in 1907, Ebenezer Cunningham, in 1914 and 1915, and Leigh Page, in 1914, using a multidirectional energetic LeSage-type ether to explain fields of force. This author is suggesting that giving the LeSage ether a magnetic force, instead of the brute force which Clerk Maxwell[28] gave it in his 1873 encyclopaedia article, would have voided his fiery universe theory. Maxwell almost killed the most adaptable ether theory ever proposed. An ether which can carry light is also able to carry energy at the speed of light. In one second of time this ether could bring in energy from a spherical volume, 186,000 miles in radius. That would explain the terrific energy in explosions. Credit S. T. Preston[11] with that thought.

Sir Edmund Whittaker, Sir J. J. Thomson, Ebenezer Cunningham, and Leigh Page were intrigued with the idea that a radiant ether, moving through the influence of protons and electrons at the speed of light, could radiate a field of force in all directions. This author is claiming that, if this radiation is equal in all directions, at the short range where the proton's force is dominant, the field of the hydrogen proton, in the hydrogen atom, will be spherical.

The area of the surface of a sphere is given by the equation, 4 pi r^2. 2 pi times 2 pi would be equivalent to multiplying the latitude by the longitude to specify every point on the surface of the earth. Dividing the energy in that field by 2 pi, as the physicists do in quantum theory, would restrict our attention to the energy in the 2 pi length of the path of an orbitting hydrogen electron.

The r in the field equation is expressed in wave lengths. The r^2 in the equation above, squares the length of the electron's orbital radius to give the number of units in each orbit and divides the 3.4 ev of angular velocity in each orbital level to give the angular velocity in each unit. The inverse square law will also derive from the r^2 above.

For the division of the energy, h, by 2 pi, see Isaac Isamov, "Understanding Physics, Volume III, pp. 73-77." The orbits of the electrons follow circular paths on the surface of a series of larger and larger spheres. The plane of the first orbit and the planes of the succeeding orbits should all lie in the same disc-like plane with the proton in the center of all of them. In effect, we would have standing waves ready to be occupied.

Sir J. J. Thomson[3] in 1893 made the useful suggestion that there was a fan-like mechanism within a magnet that could send two lines of magnetic force in opposite directions at the speed of light. If rotating waves are moving through each other in opposite directions it would seem that J. J. Thomson is describing a magnetic double helix such as was discussed in this author's 1992 book, "The Magnetic Double Helix." It was mentioned there that there could also be an electric double helix to tie the electron to the proton. The double helix gives rigidity to bonding because every double bond pulls in opposite directions, at the same time, thereby protecting against both stretching and compression. The rigidity also aids the contractions.

There is a de facto creation of momentum at each orbital level. At the first level the laws of momentum decree that the electron's captive energy is divided into 4 units of 3.4 electron volts each, for a total of 13.6 ev of captive energy. There are four units of distance to go with the four units of angular velocity. The field equation takes care of all the proportions. If the electron loses even a small part of that voltage, the atom can become an ion and the electron can fly off into space. That is why that original energy is called captive.

In order for the electron to pull itself away from the proton an additional distance of one wavelength, the electron must absorb a quantity of free energy equivalent to 3/4 of the value of its captive energy in that level. The second orbit has the same angular velocity, 3.4 ev, as the first, but it has seven units of that angular velocity instead of four. It needs the extra energy because it has three more units of distance to travel at the second orbital level, It would then contain three units of free energy and four units of captive energy.

At each hydrogen electron orbital level, where there is free energy that has been absorbed previously, this energy may be radiated as a photon of light. The 10.2 ev of free energy may be radiated as a photon of light if the electron returns to the first orbital level from the second level. The electron cannot bypass the second level in order to occupy a higher level. Each level builds on the level below it.

The r in the field theory is expressed in wavelengths. The r^2 in the field equation above, squares the length of the electron's orbital radius to give the number of units in each orbit, and divides the 3.4 ev of angular velocity in each orbital level, to give the angular velocity in each unit. The number of units of angular velocity at each orbital level, and its angular velocity, starting with the third level, will be, 9 X .377 ev, 16 X .2125 ev, 25 X .136 ev, 36 X .0944 ev,

49 X .9693 ev, 64 X .0531 ev, etc. Four will be subtracted from the number of units in each orbit that can enter into the radiation of a light photon. Only the free energy can be radiated.

The radiant energy of the sun, and the radiant energy of the hydrogen atom's proton, control orbits with very similar types of forces in their fields. The values of the angular velocity of the planets are much more approximate than the exact values of the angular velocity of the electrons. In the case of the planets, they have been subjected to many collisions since they started forming. More planets have disappeared than have survived. See the new Bode's law in the chapter on the planetary orbits.

The captive energy plus the radiated free energy.

1. 4 X (1/4 X 13.6 ev)
2. 4 X (1/4 X 13.6 ev) + 3 X (1/4 X 13.6 ev)
3. 4 X (1/9 X 3.4 ev) + 5 X (1/9 X 3.4 ev)
4. 4 X (1/16 X 3.4 ev) + 12 X (1/16 X 3.4 ev)
5. 4 X (1/25 X 3.4 ev) + 21 X (1/25 X 3.4 ev)
6. 4 X (1/36 X 3.4 ev) + 32 X (1/36 X 3.4 ev), etc.

Chapter 2

Inverse Square Momentum

The atoms in this keyboard are moving with the earth at a certain speed which we will label, V. These atoms are held together by bonding waves which move at the speed of light which we will label, C. What would be the speed of the bonding waves relative to the speed of the atoms?

Relativity theory will say that the waves will move at the speed of light, C, relative to the atoms, regardless of the speed of the atoms or the direction of their travel. There are interesting consequences, however, if we figure the relative speeds and wavelengths to be (C – V) and (C + V), in the two directions along the line of momentum, and if we allow the bonds to obey the inverse-square law.

A bond will be stretched in the forward direction and compressed in the rearward direction because the bonding wave follows in the first case and meets in the second case. The length of the waves to the rear can be written as (C – V) and the length of the waves to the front can be written as (C + V). If these bonds obey Maxwell's law for electromagnetic waves, the inverse-square law, the shorter waves to the rear will pull more strongly than the waves to the front. This would seem to offer a cause for momentum at less than the speed of light.

It seemed that this was a problem that could be solved. One should only need some simple algebra. If momentum does lie in the pull of bonding waves, the algebra should show that the excess pull, in the direction of motion, varies directly with the speed of the system. Twice as much pull should give twice as much speed. This should hold true except when speeds approach the speed of light. Research has shown that acceleration gets very difficult when speeds get that high.

The strength of pull varies inversely with the square of the distance a wave travels. This can be written as $1/d^2$. The speed of the bonded material, the magnet, or the system, is labeled, V. The speed, V, is always a fraction of the speed of light, C. When C is given the value, 1, the speed of the system can be written as V/C, which can be a decimal fraction of l. Then, C -V becomes 1 - V/C and C + V becomes l + V/C. In any one unit of time, these speeds can be translated into the relative distances traveled by the bonding waves.

We should calculate the pull from the front and the pull from the rear and subtract the later from the former. A table will be worked out for ten values of V/C, starting with V/C = .0001 and for another ten values of V/C, starting with V/C = .1. The pull represents the amount of energy that would be needed to create the speed from scratch. Remember that these are only proportions.

In order that there will be no doubt as to the method of calculating the proportions, the first value for V/C will be worked out in detail. Starting with V/C = .0001, we proceed as follows. A pocket calculator is sufficient.

1 - V/C = 1 – .0001 or .9999.
1 + V/C = 1 + .0001 or 1.0001.
$1 / (.9999)^2 - 1 / (1.0001)^2$
1 / .9998 – 1 / 1.0002
1.0002 - .9998 = .0004 = pull at speed, (V/C =.0001).

V/C		V/C	
V/C = .0001	.0004	V/C = .1	-.408121
V/C = .0002	.0008	V/C = .2	-.868055
V/C = .0003	.0012	V/C = .3	1.449100
V/C = .0004	.0016	V/C = .4	2.267573
V/C = .0005	.0020	V/C = .5	3.5555556
V/C = .0006	.0024	V/C = .6	5.859375
V/C = .0007	.0028	V/C = .7	10.76509
V/C = .0008	.0032	V/C = .8	24.691358
V/C = .0009	.0036	V/C = .9	99.722992
V/C = .0010	.0040	V/C = .99	9999.7475
V/C = .01	.040008	V/C = .999	999999.

The formula, used above, to find the amount of energy needed to create the stated speeds, is as follows,

$1 / (1 - V/C)^2 - 1 / (1 + V/C)^2$.

The formula used in relativity theory to determine what they call an increase in weight, or mass, is as follows,

$1 / (1 - V^2/C^2)^{1/2}$.

The energy needed to create a comparable amount of speed, at the higher speeds, becomes much less in relativity theory, as can he shown by the table below.

V/C		V/C	
V/C = .001	1.0000005	V/C = .9	2.2294
V/C = .002	1.000002	V/C = .99	7.0881
V/C = .003	1.0000045	V/C = .999	22.3662
V/C = .004	1.000008	V/C = .9999	70.7124
V/C = .005	1.0000125	V/C = .99999	223.6073
V/C = .006	1.000018	V/C = .999999	707.1067
V/C = .007	1.0000245	V/C = .9999999	2236.068
V/C = .008	1.000032	V/C = .99999999	7072.
V/C = .009	1.0000405	V/C = .999999999	22360.
V/C = .010	1.00005	V/C = .999999999+	40000.

The last figure, showing the mass increasing 40,000 times, is taken from the well known book by Martin Gardner[6], "Relativity for the Millions." The indefinite, +, was used by Gardner, page 69.

There are ways to estimate the input of energy into an accelerator. From that, relativity theory can use a chart, such as the one above, to estimate the amount of speed achieved. The (C - V) and (C + V) theory or the relativity theory, either one could supply the great increase in resistance to acceleration which has been observed in accelerator experiments. Note that the table for the mass increase given by Einstein's theory does not show the mass increasing in proportion to the speed. In the table using (C – V) and (C + V), the proportion is very faithful until the speeds approach the speed of light.

A mass increase at high speeds has long been cited as a major support for relativity theory. The increasing resistance to acceleration gives stronger support to a (C - V) and (C + V) theory and does so without logic-defying postulates such as flexible time and bent space. This theory of momentum requires no additional postulates of any kind. Newton's inverse-square law as adapted by Maxwell to electromagnetic radiation is the only support needed. The rest is the result of following mathematical and geometrical relationships and common assumptions of electron activity.

Alfred North Whitehead[7] seemed to believe that the first law of motion, as created by Galileo, was one of the great advances in science and was more than a mere statement of belief. Newton is quoted as saying that, "Every body continues in its state of rest or of uniform motion in a straight line except so far as it is compelled by impressed force to change that state." Whitehead explained that they had been looking for 2000 years for the cause of momentum. This search had blocked the progress of science for all that time. He said that we now knew that it was just the natural state of matter. The theory in this book seems to be committing the crime of rewriting Galileo's first law of motion as well as putting travel time into the inverse-square law.

There is a historical precedent for a theory which would put the energy of momentum in the ether. In 1893, J. J. Thomson[3] proposed that the ether was a storehouse for the energy of momentum. Also, there is experimental proof that magnetic waves can carry momentum. The proof lies as close as the nearest bar magnet. The waves from a bar magnet can impart momentum to a piece of iron. There is also the evidence that can be gathered from the explanation of the minus 4 in the Balmer formula. The minus 4 in the Balmer puzzle is the 4 units of momentum which is never radiated. This is starter momentum for captive energy which can count in the necessity for a full orbit but must always drop to a lower level with the electron instead of being radiated.

Chapter 3

Balmer Formulas.

The Balmer constant, .000036456 cm, is symbolized B. The Rydberg constant, 109737.31 cm, is R. The Balmer formula is, wavelength = B $n^2/(n^2 - 4)$, where n can have successive whole number values, beginning with 3. If n = 3, then B 9/(9 – 4) = .000065621 cm and that will be the wavelength of the third level Balmer line in the hydrogen spectrum.

The Rydberg formula for solving for the Balmer wavelengths is 1/w = R ($1/2^2$ - $1/n^2$). When n = 3, that reads R (.25 – .111111) and that equals, R (.1388888). The reciprocals of R and .1388888 will be .000009113 and 7.2 and the product of those factors will be .000065614 cm, which is the wavelength of the third level line of the Balmer spectrum.

There is a new way to solve with a wave formula, a way that removes that minus sign in the formula or explains why it is there.

4/Balmer constant = Rydberg constant = R = 4/B and
4/.000036456 cm = 109737.31 cm = R = 4/B.

When the formula is adjusted to the Lyman series of wavelengths, and at the same time, n = 3, the formula will read in this way, l/w = 4/B ($1/1^2$ – $1/3^2$). Multiplying within the parenthesis, 1/B (4 - 4/9) reduces to, 1/B (3 + 5/9). That can be simplified to l/B (3.5555555). The reciprocals of the two factors, are .000036456 cm and .28125. The product of the two is .000010253 cm and that is the third level Lyman wavelength.

The formula for arriving at the second level Rydberg wavelength is 1/w = 4/B ($1/l^2$ - $1/2^2$). That translates to 1/B (4/1 – 4/4),and that shortens to l/B (3). The reciprocals of these two factors are .000036456 and 1/3. Multiply the two factors to get the second level Lyman wavelength, which is .000012152 cm.

There is a Paschen series in the infrared range. One equation for that series is, when n = 4, 1/w = R ($1/3^2$ – $1/4^2$). This equation can be changed to, l/w = 4/B (1/9 – 1/16). Multiplying within the parenthesis by 4, gives us 1/B (4/9 – 4/16) and that simplifies to 1/B (.444444 – .25), and then to l/B (.194444). The reciprocals of the two factors are .000036456 cm and 5.14285 and multiplication reveals that the wavelength is .000187488 cm.

Some years after Balmer created his formula, Johannes Rydberg altered that formula to what was termed, "a more convenient form." By some legitimate algebra the formula was twisted around to its present form, 1/w = R (1/4 – l/n^2). We see how it was more convenient when they changed the 1/4 to $1/2^2$. They had to put the formulas into a reciprocal form in order to get the following series of formulas. In 1,2,3,4,5, order the following are for the Lyman series, the Balmer series, the Paschen series, the Brackett series, and the Pfund series.

1/wavelength = R $(1/1^2 - 1/n^2)$ when n = digit, 2 or more.
1/wavelength = R $(1/2^2 - 1/n^2)$ when n = digit, 3 or more.
1/wavelength = R $(1/3^2 - 1/n^2)$ when n = digit, 4 or more.
1/wavelength = R $(1/4^2 - 1/n^2)$ when n = digit, 5 or more.
1/wavelength = R $(1/5^2 - 1/n^2)$ when n = digit, 6 or more.
In all of these the R can be changed to 4/B.

Gerhard Herzberg's book[44], "Atomic Spectra and Atomic Structure," lists 21 hydrogen wavelengths for the above 5 series. Starting with the Lyman series, the counts for these series were 5, 8, 5, 2, and l.

Sir James Jeans in his book[45], "The New Background of Science," saw great significance in the accuracy of these wavelengths, an accuracy of over 1 part in a hundred thousand. In 1934, just a few years after the major work on quantum theory was completed, Jeans was saying that, while we could not know anything for certain, a theory which predicted 20 lines so accurately could not be a mere coincidence. The odds in favor of the predictions of the new quantum theory are millions to one. Jeans says that the quantum scheme for the origins of the hydrogen spectrum is true in its numerical essentials.

This author claims the same millions-to-one accuracy for he has the same predictions for the same wavelengths. This author, however, needed nothing but common sense arguments and simple algebra. He did not need to abolish the relative speed of light, invent a fourth dimension, en1ist the magic of the square root of minus one, and use the mystical notions of Hamiltonian mathematics, for this is a common-sense universe.

Chapter 4

Momentum in Quantum Orbits

$E = (6.626 \times 10^{-34}$ J/s$) \times (3 \times 10^{8}$ m/s$)$
1.9878×10^{-25} J/s. / 656×10^{-9} m $= 3.03 \times 10^{-19}$ J/s.
E (ev) $= 3.03 \times 10^{-19}$ J/s $\times$ 1 ev / 1.6022^{-19} J/s $= 1.89$ ev.

When the voltage, 1.89 ev, is right for the third level Balmer orbit, the following theory, which leads to the same result, may also be right.

Several years ago, while studying several aspects of the inverse square law, this author discovered some very strong evidence which says that there is a 4 to 1 ratio between the voltage needed for creating momentum and the voltage of the momentum. This was published in this author's 1992 book, "The Magnetic Double Helix." Because that is part of the evidence for the creation of momentum for the quantum orbits, that chapter on momentum is chapter 2 in this book. Chapter 5, which follows, will explain momentum as it applies to the orbits of the planets.

Part of the evidence for momentum goes back to Newton. His gravitational formula mentions two masses which pull on each other. Each mass contributes to the total pull. There must be two distances also, a d_1 and a d_2. Probably it hasn't occurred to scientists that there could be two distances, different from each other. We should, however, consider the geometry of the situation as it might occur in a crystal.

If two radiating bodies have a motion in the same direction with one following the other in the line of motion, while some attractive force is passing between them, they will be in the situation where a wave force from the front will meet a body coming from the rear and a wave force from the rear will overtake the body to the front. If the speed of the waves is C, and the speed of the bodies is V, the wave to the rear will travel a distance C – V, while the wave to the front will travel a distance C + V. The inverse square law says that the waves get weaker with distance. The pulls from the front are shorter and pull more strongly. The difference between the two pulls is momentum. The momentum balances an energy equation. The difference between C - V and C + V becomes the energy of momentum.

The energy of each wave will diminish according to the inverse square law, so it is easy to figure the relative energy left in each wave at the time the waves reach their targets. It turns out that the difference between the two pulls is exactly four times the speed of the momentum. The exactness begins to disappear as speeds get very high. At 1/100 of the speed of light there is some inexactness showing at the seventh decimal point.

At close to the speed of light, the amount of pull needed to create the speed of momentum increases more rapidly than it does with Einstein's formula for what he called an increase in mass. That doesn't mean that atomic energy is not real. It just means that the energy of momentum has a reality of its own and shows it in this way. It may also be that large stores of momentum are stored in the nucleus. Gamma rays may be a release of momentum.

Chapter 2 gives a mathematical reason for believing that the value of straight-line momentum is created by applying 4 units of energy for each one unit of momentum. We find that giving an electron 4 units of the energy of angular velocity creates the momentum of the continuous orbiting in atoms. There is, however, another requirement for the creation of momentum in orbits. A full orbit is necessary. The momentum must be paid for by enough energy to carry the electron around a complete circumference. There is no special notation for momentum.

There are several things special about the first hydrogen orbit. For one thing, this is the place where angular momentum is first formed. At the distance of the first orbit the proton radiates a pull of 13.6 ev. The electron resists that pull with 13.6 ev of its own. The electron distributes that energy evenly around its orbit in 4 units of 3.4 ev each. The unit system was devised in the early years of the study of the planets when the first planet's energy was divided into 4 units as compared to the earth's possession of 10 units of energy. Four units in an orbit is the minimum distribution in the creation of momentum. This is a full 4-unit orbit. The units of angular momentum were judged by the measurements of the various orbital radii.

That 13.6 ev of angular momentum is called captive energy because the electron would cease to have an orbit if the electron loses any part of it. The atom would become an ion. The captive energy figures in the field equation. The first 4 units of an orbit are all captive energy and those orbits set the value for all the units of angular velocity in an orbit. The electron can advance to higher orbits, various wavelengths of distance from the proton in the hydrogen atom, by absorbing the proper amount of free energy.

The electron must absorb 10.2 ev of free energy in order to pull itself a second wave length of distance away from the proton to the second level orbit. There, the electron will have 4 units of captive energy and 3 units of free energy. Four units in the first orbit and 7 units in the second. This agrees with what we know about the first and second levels in the planetary orbits. The angular velocity is still 3.4 ev at the second level but that orbit is 3 units longer and needs 3 more units of energy. The first 2 levels each have a total of 13.6 ev of captive energy.

The difference between the Leman series of orbits and the Balmer series is that the 10.2 ev of free energy that was absorbed by the electron for the second orbit cannot be radiated by the electron until the electron returns to the first orbit.

There is a similar stopping point for other series and there is always the proviso that only free energy can be radiated. There is also the fact that 10.2 ev of the proton's pull is tied up in the second orbit and the higher orbits are limited to 3.4 ev, that is, except for the Lyman series.

The electron uses 1.89+ ev in pulling against the pull of the proton to that third orbital level. Subtracting 1.89 ev from 3.4 ev equals 1.51 ev. That would be 4 x .377 ev. That is all the captive energy left in that third level orbit. That is the starter for that level's momentum. If we add 1.89 of free energy, we would add 5 x .377 ev of free energy, to the 4 units of captive energy and will have a full orbit of 9 units of .377 ev each for a total of 9 units of angular velocity.

In order to rise from its position at the 2nd orbital level to the 4th orbital level the electron must use 2.55 ev of its captive energy. 3.4 ev minus 2.55 ev is .85 ev. That means that there would be a residue of .85 ev of captive energy for starter momentum for the 4th level momentum. That residue, when divided, would give 4 units of .2125 ev of captive energy. That plus 12 units of free energy would make a total of 16 units of .2125 ev each. That would be a full orbit for the purpose of momentum. That would be 16/16 of 3.4 ev.

In order to rise from the 2nd orbital level to the 5th level the electron must use up 2.856 of its captive energy. .544 ev of captive energy would be left to form 4 units of .136 ev of starter momentum. Adding 21 units of .136 ev would give us 25/25 of 3.4 ev and that would be a full orbit for the purpose of creating momentum.

In order to rise from the third level to the fourth level we would subtract .85 ev from 1.51+ ev and arrive at the figure .66 ev as the cost in volts. Also, 2.55 ev – 1.89 ev = .66 ev. That is the cost of the rise from the 3rd level to the 4th level.

At each level the 4 units of starter momentum is the energy in that orbit which cannot be radiated. The energy which can be radiated at each level is the free energy which had been added to those orbits in order to make them full orbits. The number of units of free energy in each orbit can be listed as follows, 0, 3, 5, 12, 21, 32, 45, 60, etc. The free energy added at lower levels can be carried to higher levels. The shrinkage in the captive energy at each level is replaced with free energy until the orbital energy is almost all free energy. At the 10th level the captive energy would be 4% and the free energy would he 96%.

3 x 3.4 ev	=	10.2	ev.	Voltages for first 6
3 + (5/9 x 3.4 ev)	=	12.08+	ev.	lines of the Lyman
3 + (12/16 x 3.4 ev)	=	12.75	ev.	series of the hydrogen
3 + (21/25 x 3.4 ev)	=	13.056	ev.	wavelengths.
3 + (32/36 x 3.4 ev)	=	13.22+	ev.	
3 + (45/49 x 3.4 ev)	=	13.32+	ev.	

5/9 x 3.4 ev = (5 x .377+ ev) = 1.89+ ev Voltages for

12/16 x 3.4 ev = (12 x .2125 ev)	= 2.55 ev		the first five
21/25 x 3.4 ev = (21 x .136 ev)	= 2.856 ev		lines of the
32/36 x 3.4 ev = (32 x .0944 ev)	= 3.022 ev		Balmer series
45/49 x 3.4 ev = (45 x .069 ev)	= 3.12+ ev		

In every orbit there will be 4 units of captive energy which can function as the starter energy for angular momentum for that orbit. In every orbit the 4 units of starter energy is the residue of captive energy which is left after the electron has spent most of its energy in moving to a higher orbit. At each orbit that residue becomes smaller as the distance becomes greater. That starter energy will be, beginning with the third orbit, 4/9ths of 3.4 ev, 4/16ths of 3.4 ev, 4/25ths of 3.4 ev, 4/36ths of 3.4 ev, 4/49ths of 3.4 ev, and 4/64ths of 3.4 ev, etc. That is the captive energy at each of those levels.

Some Background

Every point in free space is a radiant center. As the earth follows its corkscrew path through space, its human passengers sample the view of the universe in all directions. The astronomers tell us that they see light coming from billions of stars from billions of different directions. They see all of them through the small eyepiece of a telescope. The light waves must be terribly crowded as they enter the small pupil of the eye.

During billions of years of travel, a particular bundle of light waves, enough to influence an electron orbit in the film in a camera, must pass through billions of billions of centers of radiation, each as small as the eye's pupil. At each individual center that the above light wave must pass through, they must pass through millions of other light waves coming from and going out in millions of other directions. There is one feature of this light travel that decreases the crowded condition. The light that passes in one second is spread out over a path 186,000 miles long.

The gas laws would say that the mean free path of disturbances in the ether, like light waves or other fields, would increase as the cube of the decrease in size. If a gas particle's decrease in size is to 1/10th, the mean increase in free path would be to a thousand times. If a particle size dropped to 10×10^{-100} the mean free path would increase to 10×10^{300}. See Chapter 8 of this book.

This has been leading up to a discussion of the inverse square law. The total pull of the radiation from a radiating body, as it is measured from any direction, can be averaged so as to appear to come from the center of that body. Radiation expands in all directions, weakens as it spreads over the larger surface, while it is moving at the speed of light in all directions from the center. In other words, radiation is a spherical phenomena.

The formula for calculating the area of the surface of a sphere is 4 pi r^2. The radiant forces which control the orbits are divided along the length of a 2 pi circumference, and weaken inversely with the square of the distance to a point on that circumference. The electromagnetic field which extends from the hydrogen proton in all directions, will extend in all those directions even if the electron is not there.

Chapter 5

Momentum in Planetary Orbits

Johannes Kepler (1571-1630), was one of the pioneers in the study of the mathematical relationships of the various planets to each other and to the planet earth[43]. He did this using astronomical units in which comparisons were to values relating to the Earth's values, as its base. The total value of the 10 units of angular velocity in the earth's orbit is 10 times .1 A.U. or 1 A.U.

One of Kepler's laws was that every planet's orbit was an ellipse with the sun as one focus. Before Bode's time and for a long time afterward circles were preferred. Circles had the blessing of religion.

Another of his laws pointed out that, on a map, a straight line drawn from a planet to the sun would pass over equal areas during equal times. On a day to day basis a long line would move slowly and a short line would move rapidly.

Kepler's third law was that the square of the length of a planet's yearly orbit will be equal to the cube of the length of its mean distances from the sun. His third law is summed up in the table below.

Kepler's Harmonic Law

Planet	Average Distance From Sun	Period in Years	Cube of the Distance	Square of the period
Mercury	.387	.241	.058	.058
Venus	.723	.615	.378	.378
Earth	1	1	1	1
Mars	1.524	1.881	3.540	3.538
Jupiter	5.203	11.862	140.707	140.851
Saturn	9.539	29.458	867.977	867.774

The following is this author's suggestion for computing the length of a planet's year.

Mercury	$.387^{1/2}$ =	.6221	x	(4 x .097)	=	.241
Venus	$.723^{1/2}$ =	.8503	x	(7 x .1033)	=	.615
Earth	$1.00^{1/2}$ =	1.	x	(10 x .1)	=	1.0
Mars	$1.524^{1/2}$ =	1.234	x	(16 x .09525)	=	1.881
Jupiter	$5.203^{1/2}$ =	2.281	x	(49 x .10618)	=	11.868
Saturn	$9.539^{1/2}$ =	3.089	x	(100 x .09539)	=	29.461
Uranus	$19.19^{1/2}$ =	4.381	x	(196 x .097908)	=	84.064
Neptune	$30.07^{1/2}$ =	5.4836	x	(289 x .10405)	=	164.892

Pluto $39.52^{1/2}$ = 6.296 x (400 x .0988) = 248.442

The square root of the planet's average distance from the sun, times the number of units of momentum in that orbit, times the average angular velocity in each unit, gives the length of a planet's year when the earth's year is 1 A.U.

There is a short-cut for figuring the length of the planet's year in A. U. years. The square root of the planet's distance from the sun, times the planet's distance from the sun gives the length of the planet's year. For instance, .387 ½ x .387 = .241, in A.U. values. This only works when the distances are the average distances over a year's time.

In elliptic orbits the number of units of angular velocity is the same all year. When the radius increases, it is automatic that the square root of that length increases also. It is the length of the unit over which the energy is expended that controls the speed over the course of the year.

If we try to compare the values in the planetary orbits with the values in the hydrogen orbits we must make allowance for the fact that there are no spectrum lines involved. The Lyman series would be where the comparison would be. The comparison fails when the Earth's orbit is found to have ten units of momentum. This was probably caused by collisions between the original circular orbits having been arranged by momentum rather than by mass. As the disc shrunk and flattened the gravitational forces could cause collisions which could result in ellipticities.

Bodes Law

In 1766 Johann Titius subtracted 4 from the value of each planetary distance from the sun and found a series of remainders which doubled from planet to planet. Johann Bode, who was an astronomical publisher, gave the idea a lot of publicity. So much so that it finally became known as Bode's law. Only six planets were known in Bode's time. When the seventh planet was discovered the orbital axis increase was close to being double there also. With the last two planets, Neptune and Pluto, howeverer, the doubled increase of the remainder did not occur.

Bode's Law

	A		B		C	D
Mercury	.4	+	.0	=	.4	.387
Venus	.4	+	.3	=	.7	.72
Earth	.4	+	.6	=	1.0	1.0
Mars	.4	+	1.2	=	1.6	1.52
Gap	2.4					
Jupiter	.4	+	4.8	=	5.2	5.2

Saturn	.4	+	9.6	=	10.0	9.54
Uranus	.4	+	19.2	=	19.6	19.19
Neptune	.4	+	38.4	=	38.8	30.7
Pluto	.4	+	76.8	=	77.2	39.52

A + B equals C. In column D you find the measured distances from the planets to the sun, averaged over a year's time. Due to the elliptical nature of the planetary orbits, the distance to the sun is constantly changing. Entering 2.4 A.U. for the missing planet, the figures in column B double at a constant rate all the way to Pluto. The actual value in column D is quite close to the value in column C until we arrive at Neptune and Pluto. For Neptune, the Bode theory would call for the actual figure for the planet's average orbital radius to be 8. A.U. higher than it is. For Pluto the planet's average orbital radius should he about 37.68 A.U. higher than it is.

Bode's Law Revised by Hagemier in 1995

	A	B		C		D		E
Mercury	(4 + 0)	0.097	x	4				.387
Venus	(4 + 3)	0.1033	x	7	x	7/7	=	.723
Earth	(4 + 6)	0.1	x	10	x	10/10	=	1.00
Mars	(4 + 12)	0.0953	x	16	x	16/16	=	1.524
Gap	(4 + 21)							
Gap	(4 + 32)							
Jupiter	(4 + 45)	0.106	x	49	x	49/49	=	5.203
Gap	(4 + 60)							
Gap	(4 + 77)							
Saturn	(4 + 96)	0.0954	x	100	x	100/100	=	9.54
Gap	(4 + 117)							
Gap	(4 + 140)							
Gap	(4 + 165)							
Uranus	(4 + 192)	0.098	x	196	x	196/196	=	19.19
Gap	(4 + 221)							
Gap	(4 + 252)							
Neptune	(4 + 285)	0.104	x	289	x	289/289	=	30.07
Gap	(4 + 320)							
Gap	(4 + 357)							
Pluto	(4 + 396)	0.0988	x	400	x	400/400	=	39.52

Column A gives the total number of units of angular velocity in a planet's orbit. Column B gives the total amount of angular velocity in each unit, in A. U

s. Column C gives the number of units of angular velocity gathered together to form the planet"s momentum. Column D is the equalizer factor. That means that the factor that lessens the Sun's power to pull on the planet's orbit (the inverse-square effect and the increase in distance), is the same factor that multiplies the ability of the planet's angular velocity to resist that force. We should increase the planet's effective energy by the same amount. Column E shows the planet's measured distance from the sun, averaged over a year's time. That is the point where the planet's angular velocity exactly balances the pull of the Sun.

Many scientists thought that there might be a planet between Mars and Juniper. Even Kepler had discussed that. They were disappointed in that they could find nothing that was worthy of the name, planet. The new theory shows many vacancies. In the early stages of planet formation the planets had to carve out wide bands of space to move in. Collisions could cause the ellipticity of the orbits. Elliptic orbits require more room. Additional angular velocity and longer orbits need more room.

There are 4 and 7 units of momentum in the first and second planetary orbits for the same reason that it is 4 and 7 in the quantum orbits. The pull of the sun is divided by 4 because the initiation of momentum requires 4 units of angular velocity in that first orbit. It wouldn't matter how much the angular velocity is. The main point is that it should be divided evenly. This sets the values for all the astronomical units for all the orbits. In addition to needing 4 units of momentum there is also the necessity for a full orbit. Venus has a longer radius and orbital circumference, therefore Venus needs 3 more units of angular velocity for a total of 7. Mercury had 4 units.

The captive energy in the quantum orbits will be given a different title in the planetary orbits. Call it positional energy. In the quantum orbits the captive energy was also positional energy. Those positions were spots where the starter energy was just waiting for some free energy to come along to start a new orbit. In each case the electron used all the positional energy that was available at that distance.

The fact that a tiny system in an atom and a mighty solar system are organized on such a similar plan must mean that the solar system, at or before the time of planet formation, had already been organized according to the laws of momentum into a disc-like structure. The momentum in our solar system must have already sorted itself while it was gas and dust. If the growing planets were too crowded some of the orbits may have merged. The gaps may have formed because of that very reason. The longer orbits would need more room for straying. The ellipticity of an orbit could also figure into the need for a wide orbital band.

In the quantum orbits, the third orbit is a 9 unit orbit. The Earth's orbit is a third orbit but it is an 10 unit orbit. It looks like Venus would have been crowding the Earth's 9 unit orbit with its 7 unit orbit which was already over loaded with its .723 A.U. of angular velocity. The Earth's 10 unit orbit had the average .1 of an A.U. of angular velocity for each unit.

Chapter 6

The Magnetic Double Helix

James Watson[1], in his book, "The Double Helix", says that the discovery of the DNA structure was the result of model building. In that book, Francis Crick explains to Watson that Linus Pauling's discovery of the alpha helix was the product of common sense and not the product of complicated mathematical reasoning. Pauling needed very few equations to elaborate his arguments for the structure of the alpha helix. Crick and Watson set out to build models and solve the structure of DNA, using the same methods. They built a model in which a great many facts fit together so neatly that they felt that it almost had to be right. Watson emphasizes that model building is a serious approach to science.

This book contains the description of another double helix. There may be a relationship between the two helices for the magnetic double helix would exist in the bonds which hold atoms together. The helical twist of the bond may carry over to the twist of the DNA. The interlocking mathematical evidence goes far beyond any possibility that its existence is due to contrived or accidental coincidence. Models could be made from the drawings for the relationships are mechanical, mathematical, and precise.

A model is more than a theory. A model supports itself from many directions. Like a jigsaw puzzle, a model proves itself by the way it fits together. This is more than two or three coincidences. A score or more of facets of physical theory are related to each other by the mathematics of (C – V) and (C + V). The cross relationships are immense because a factorial quantity is much greater than the sum of its parts. A factorial number is used in the calculation of mathematical odds. 1 x 2 x 3 x 4 = 24 = 4!. There would be 24 ways that 4 facts could support each other if they can all be mathematically related to each other. If, for instance, we have 10 significant experimental facts to support a theory, the factorial odds would be 3,628,000 to 1. It is a practical certainty that the magnetic double helix does exist.

This author feels free to criticize relativity theory because he has found what the physicists looked for and never found. This author found the common-sense mathematics that can explain the Michelson-Morley experiment. Lorentz was wrong in using the ratio of two contraction as one contraction. We need two separate contractions. One lateral to the line of motion and one in the line of motion.

This author feels free to criticize quantum theory for a similar reason. He found a common-sense mathematics that can be used to justify the identical wavelengths that are being used with such spectacular success by chemists and physicists today.

He has also found a common-sense solution for the creation of momentum at less than the speed of light. Knowing how momentum is created can tell us how momentum is spent in creating light waves. The Balmer Formula is also explained by the use of simple mathematics. Bodes Law for the planetary orbits needed some correction when they reached the last two orbits, those of Neptune and Pluto. The theory of momentum that explains quantum orbits will also explain the planetary orbits. This author believes that the physicists should start using the mathematics which Einstein abolished, (C – V) and (C + V). The physicists should try to wean themselves from their recondite Hamiltonian mathematics and stop trying to put skids under the rule of reason.

In 1934, Sir James Jeans[45] gave no sympathy to the complaint that the new science was contrary to common sense. He said that the age of common-sense science was past; new and unfamiliar concepts have been found to be necessary. Sir Jeans was agreeing with the pioneers in quantum theory and quantum mechanics, that common sense was not capable of explaining the mysterious meaning of the Balmer formula. He was wrong. Einstein pleaded great necessity when he introduced his radical departures from common sense. He was wrong.

In his 1905 theory, Einstein was proposing that a fourth dimension should be added to the three which we already had. He also seemed to be saying the speed of light was constant relative to any other speed. This meant that no speed could be relative to the speed of light. Common sense would say that all speeds are relative to each other. Its not a one-way street.

Most of the mathematics of relativity theory is very abstruse, but some of it is simply magical. There is the case of the square root of minus one, usually written as little i. In using i, it is as if one crosses his fingers behind his back, writes little i and jumps from a minus sign to a plus sign, or writes little i and jumps from three dimensions to four dimensions. This is only a small exaggeration.

In the early twentieth century a great many physicists had relegated the Michelson-Morley experiment to the history books and were then concerned that another puzzle demanded a solution. That was the Balmer formula for determining wavelengths in the hydrogen spectrum. This problem and this author's solution will be covered in the chapters 2, 3, 4, and 5. The Balmer formula was very important. It had some of the aspects of a Rosetta Stone in that it gave us our first hints on how the waves of the spectrum are related to each other.

The Balmer formula, however, had failed to yield the secret of how a minus 4 fitted into a common-sense explanation of the formation of the lines in the hydrogen spectrum. Guy Murchie[43] in, "Music of the Spheres" tells the following story about Paul Dirac. In 1927, Dirac found the magic combination of h, 2pi, and little i. In one lightning-like flash of revelation that was, "the most exciting moment" of his life, he justified the formulas for the wavelengths of the hydrogen spectrum. Dirac also adopted the p's and q's of Hamiltonian mathematics. Along with little i, they were a new kind of number that could stand for abstract quantities such as momentum and position. They were like operators such as +, –, or x. p x q would not necessarily be equal to q x p.

Einstein used the Lorentz transformation equations as the basis for his special theory of relativity. There was a time discrepancy in Newton's inverse-square law. This fact was discovered when the Michelson-Morley experiment failed to detect the movement of the earth through a stationary ether. When Einstein created his so-called constant speed of light, he was really abolishing the relative speed of light. (C – V) and (C + V) are the most direct ways of portraying the relative speed of light. C is used to represent any wave force which moves at the speed of light waves. V is used to represent the speed of any kind of mass or particle which moves at less than the speed of light.

When Einstein adopted the Lorentz transformations he knew that it included a fourth dimension. He was surprised at how much attention it received. If some people were going to think it was a ridiculous idea, Lorentz could share some of the blame. Lorentz may have really thought it was a ridiculous idea and may have introduced the idea as only a way of showing how far he was from solving the puzzling null results of the Michelson-Morley experiment. It is a well-known fact that, to the day he died, in 1928, Lorentz could not accept Einstein's relativity theory[17].

It is ironic that, when Einstein abolished the relative speed of light, he abolished the only mathematics that could put time into Newton's laws and explain the null results of the Michelson-Morley experiment, all by the use of plain, ordinary, and the simplest kind of algebra. Einstein abolished (C – V) and (C + V) because they were no longer needed. Using them, along with the Lorentz transformations, he could have had a double correction for that time factor which was missing in Newton's laws. When he abolished the relative speed of the waves he also abolished the use of the inverse-square law as it affects the bonds within matter.

The theory of the Magnetic Double Helix, which can also be called the (C – V) and (C + V) theory, is, for the most part, about the bonds which hold matter together. The earth pulls on the sun and the sun pulls on the earth. We were taught that in school. We should also be teaching that the bonds between all kinds of matter, in all kinds of two-war relationships, operate in two directions at once, whether they are electrical, magnetic, or gravitational.

The history of theories of electricity and magnetism is referred to in order to show that many of the ideas developed in this hook are neither radical or new. They originated in the minds of some of the most respected men in the history of science. Sir J. J. Thomson, for instance, proved the existence of the electron in

1898. Before then they did not have theories about electrons in magnetic orbits. At the time they were working, many concepts were in the future. It is necessary to blend the past with the future.

This is not the classical theory. The introduction of travel time into all the gravitational and electromagnetic interrelationships makes this a different theory. This theory retains most of classical theory, following the example set by relativity theory in this respect. Most of the changes are in the bonds that hold atoms and molecules together and apart.

This is a theory which attempts to explain a large number of commonly known magnetic phenomena as well as some that are not commonly known. There is also a description of an ether that could carry every known kind of field, electrical, magnetic, and gravitational. This ether could carry every kind of wave. This ether could be the source of all energy, including that needed for momentum at less that the speed of light. This ether could be the source of a unified field theory. Maxwell[2] said that this ether could account for the inverse-square law of gravitation if it were otherwise acceptable. The theory of the magnetic double helix could survive, however, without the theory of an ether. The well-known radiant nature of magnetic waves should be able to carry the theory when the other evidence is already overwhelming.

Sir J. J. Thomson[3] should be credited with inventing the Magnetic Double Helix. He knew that the magnetic waves which emerge from the poles of a bar magnet were rotary. He pictured forces traveling in both directions along the line of magnetic force, moving at the speed of light. He said that there should be a fan-like mechanism in the interior of a magnet which would create the waves and send them, at the speed of light, in the two directions along the line of the magnetic force. The fan-like mechanism accounts for the rotation of the waves.

This author put the picture together. The waves, rotating around their line of motion, would be helical. Any force which rotates around its line of movement will be helical. The fan-like mechanism would reside in every magnetic orbit along the long rows or short rows of magnetic atoms. The radius of rotation for the helical waves would be that of the electron orbits which create the waves. The oppositely flowing helical waves would continuously intersect with each other, and simultaneously, with the paths of the electrons which are creating the waves. The length of the waves would be the distance between their intersections with the orbitting electrons.

The double helix can account for the salient features of a bar magnet, such as, magnetic attraction, magnetic repulsion, the rotation of magnetic waves, the regularity of spacing in crystalline bonding, the demagnetizing effect outside the poles of a magnet, the drop in strength just inside the poles, the evenness of the magnetic force along the interior of the magnet, the ability of the individual atomic magnets to unite in phase relationships that result in incremental additions to the strength of the waves, the difference between the north and south poles, the illusion that the magnetic waves flow around a magnet to enter the other end, the nature of the bonding force, and the reason for the rigidity of magnetic bonding.

In addition to all the above, there is good evidence that momentum at less than the speed of light is a boot-strap type of operation in which the energy resides in the ether. The ether particles will be magnetic particles. If you don't like the word, ether, let the energy reside in oppositely flowing radiant waves. A difference in the opposing pulls of two-way forces, a difference that varies with the speed, is exactly proportional to the speed that creates that difference. When speeds approach the speed of light the proportionality rapidly disappears and the energy needed to increase the speeds mounts toward infinity.

There are also explanations for the null results of the Michelson-Morley experiment and for the null results of the Kennedy-Thorndyke experiment. There are explanations for a contraction of mass in the line of motion and there are explanations for a contraction of mass in the direction, lateral to the line of motion.

There are explanations for momentum at less than the speed of light. This is a breakthrough on a problem which has defied human thought for thousands of years. Scientists gave up on finding an answer for the creation of momentum. For several centuries momentum has been called an innate property of mass. The solution for momentum led this author to a common-sense solutions for the creation of the quantum orbits and for the creation of the planetary orbits.

There are scale drawings which illustrate how the contraction of mass in the line of motion enables three helical paths to maintain three-way intersections, through all changes of the speed in the line of motion. This is remarkable because the distance between intersections along each of the three paths must change with every change of speed in the line of motion. The intersections of the opposite waves are rigid to the point that changes are forced on the lengths of the intervals between the intersections. The three helical paths are drawn as slanting lines on flat paper. When the paper is rolled into a tube, the lines become helical. This is strong evidence for the existence of the Faraday tubes which Sir J. J. Thomson often mentioned.

The solution for the problem of the contraction of mass in the direction, lateral to the line of motion, a contraction that increases as speeds increase, was arrived at through the use of the concept of retarded action at a distance. This is the concept that when a moving source emits a wave that travels to a moving target, the path that the wave follows, will start at the past position of the source and end at the future position of the target. The length of the bond which controls the dimension, lateral to the line of motion, is rigidly controlled by quantum forces within the atoms. When the bonding force is exerted along a line that slants in the direction of movement, the lateral dimension shrinks.

It is only natural that a theory which can simplify so many other aspects of science should have some applications to quantum theory also. The double helix explains the rigid spacing along rows of atoms in crystals. The helical wave from the rear will be pulling to the rear at the same spot that a wave from the front will be pulling to the front. They will meet with the electron in a continuously simultaneous way. See the drawings in the back of the book.

This theory cannot be falsified by references to relativity theory. The proofs of that theory are not disproofs of this theory. To say so would be indulging in circular reasoning. This is an amendment to Newtonian mechanics and is meant to replace the mystical postulates of the current theory and to challenge the rigid certainty of those scientists who have promoted the uncertainty theory. This amendment involves a new formula for gravitational attraction which includes the time of travel for those waves.

Thomas S. Kuhn, a historian of science has written a very influential book called, "The Structure of Scientific Revolutions." Kuhn has popularized the use of the word, paradigm and seems to be saying that scientific revolutions are battles between paradigms. A paradigm is a set of beliefs that are tied together by a few common threads. It would be a belief system or a set of associated beliefs. These paradigms exist in many fields of organized knowledge where theories need loyalty.

The paradigm of relativity theory includes such things as the constant speed of light and the fourth dimension. This author's paradigm includes the relative speed of light and three dimensions of space. The paradigm of relativity theory combined with quantum theory has a powerful following which consists of almost all of our physic's establishment including their schools and journals. The establishment is able to suppress the publication of all kinds of revolutionary theories. This author takes their polite rejections for granted.

This book will be considered to be very dangerous. It is loaded with common sense and they have proved that common sense is not necessary and they state that this is not a common-sense universe. Kuhn says that, while circular arguments are not logical, they may be persuasive when backed by community support. A revolution can be a battle between two paradigms.

Because of the probability that this would receive a poor reception from the physic's community, this book has been written for a general readership. Anyone with a general interest in science is sure to find much that he can understand. Whether he believes may depend, to some extent, on his prejudice. In a major way, the conclusions drawn here are true to the content of the paradigm being used.

This author believes he has a well organized paradigm. Kuhn mentioned that the best feature that a paradigm could possess would be the ability to explain previously unexplained phenomena. Its great success in this endeavor is evident from an inspection of the following list. Causes have been found which can be substituted for many cases of innateness.

What follows is a large number of cases where there are no present explanations. For instance, explain momentum! Yes, there are now explanations for momentum. The reason for the minus four in the Balmer formula is another item. The reason for the rigidity of crystalline bonds makes sense. About seven different measurements in a bar magnet's strength are accounted for in one theory. The reason why Bodes law fails to predict the length of the orbits of Neptune and Pluto can be explained. An explanation for the creation of the inverse square law is quite simple. The similarity between the orbits of the planets and the Balmer electrons is explained exactly. There is a new gravitational formula which we didn't even know we needed. It includes the back and forward travel time separately. The G is taken care of by the first step in the creation of momentum. There is now a unified field for both the quantum and the planetary orbits.

Chapter 7

Some Historical Credits

There are several reasons for quoting the history of scientific ideas. In the first place science places great importance on giving credit to the originator of an idea. There is also the need of placing one's ideas in a much broader context. Scientific thinkers like to pursue ideas through the writings of many other thinkers. It is useful to one's readers if the names which one quotes are fairly well known and the sources are easily available. The originator of an idea, however, may not be well known; there is no help for that. He still should be quoted.

Rene Descartes

In 1644, Rene Descartes[9] advanced this theory. In order to account for the interactions between two magnets, the position of their poles determining whether attraction or repulsion occurs, Descartes postulated that the particle streams were aligned, that the particles had screw threads, and that the channels in the magnets were similarly threaded. It was about two hundred years later when it was discovered that the magnetic field was rotary. About a hundred years later this author had an similar idea in which two screw threads are helical fits, like in DNA. It is almost impossible to have an idea that is completely new.

George LeSage

George LeSage[10], a Swiss physicist, first proposed his gravitational ether theory in 1744, when he was a young man. He was still fighting to get it accepted when he died in 1803. He postulated an ether of very small particles which traveled in every direction at tremendous speeds. These particles could exert pressure on particles of matter and cause them to move. When two particles of matter sheltered each other from pressure in the direct line which would join them, they would be pushed toward each other.

LeSage saw that mass would have to be mostly empty space in order to account for the gravitational attraction of the mass in the interior of bodies. To LeSage, the most attractive feature of this ether was the fact that it would account for the inverse-square law of gravitation. This was due to the radiant nature of the forces. The surface area of a sphere increases by the formula 4 pi r^2. The strength of the pull of gravitation weakens with the square of the length of the radius. This is because the pull becomes spread over a larger area of surface. 2pi is the length of an orbit.

S. Tolver Preston

In 1875, S. Tolver Preston[11] had developed the ether theory of George LeSage into a full length book. Preston saw such an ether as gas in a container with the universe for its boundaries. In other words, the ether particles, in many ways, obeyed the gas laws. The ether particles, moving at the speed of light, could have, because of their extremely small size, mean free paths billions of miles long. In figuring the energy in such an ether, he computed that a quantity of matter, representing a total mass of one grain, multiplied by the square of the velocity of light, would enclose a store of energy upward of one thousand millions of foot pounds. This sounds very much like the formula, $E = MC^2$.

In Preston's case it was a measure of the energy in the grains of ether, and mass, as we know it, would only be reacting to this energy. As an example of the power in such an ether we could suggest that an explosion, lasting one second, could draw on energy from a sphere, 186,000 miles in radius.

Michael Faraday

Michael Faraday is credited with the creation of the term, "lines of force." In later days they were sometimes referred to as Faraday tubes. The tubes or lines of force were first suggested by the patterns which magnets made in a scattering of iron filings.

In 1845, Michael Faraday[12] found that the plane of polarization of a beam of light was rotated when the beam traveled through a slab of glass in a direction parallel to a magnetic field. It was also discovered that when the direction of the magnetic field changed, as when the magnet was turned end to end, the direction of the rotation of the plane of polarization was reversed. The amount of rotation increased with the strength of the magnetic field.

Sir William Thomson

In 1856, Sir W. Thomson[13], in referring to Faraday's magnetic rotation, stated, "The handedness of the rotation of the plane of polarization depends on the direction of propagation." Thomson argued that this distinguishing feature of Faraday rotation could only be explained on the assumption that the magnetic line of force corresponds to an axis of rotation of some of the material through which the light propagates. He concluded that the actual mechanical condition characterizing a region traversed by a mechanical line of force would be one where the axes of the molecular vortices were all aligned in one direction, this direction being the direction of the line of force. He seemed to have the concept that one axis of rotation served both ends of the magnet.

To illustrate the importance of the position of the observer, we can suggest that looking through the back of a glass clock would show the hands moving in a anti-clockwise direction. This last has a bearing on another problem in perception. We can cut a magnet in two and turn one piece end to end, so that a north pole is facing a north pole. They are the same in name, but, if you accept Thomson's proposal, the two fields emitted by the two north poles are now turning in opposite directions. There has been a confusion about like repulsing like and unlike attracting unlike. In a most basic way, they are no longer alike.

Clerk Maxwell

In 1861, Clerk Maxwell[14] said that the theory of molecular vortices was not merely a mechanical illustration or analogy but a theory with truth value which could only be proved wrong by experiments. He proposed innumerable small vortex tubes or filaments. The angular velocities of these would be proportional to the field intensity. Maxwell found difficulty in conceiving of vortices, side by side, and not coming into conflict. He suggested idle wheels in gear with both. This seemed to be a major sticking point with Maxwell. This would not be a problem with spiral tubes in close contact. The spirals would only have to stay synchronized. The leading edge of the wave force rotates around the spiral once every wavelength. The force will only exist at the leading edge. If the parallel tubes have their force, all in the same quadrant, at the same time, they will never come in conflict. Imagine a few barber poles standing next to each other, arranged in an orderly fashion.

George Fitzgerald

In 1885, G. Fitzgerald[15] submitted the idea that, in a magnet, we could have what he called "wheels", fastened together with elastic forces, all moving with the same orbital velocity. It could be as if they were geared together with rubber bands to turn in elastic union. Fitzgerald also suggested that the vortex filaments of the magnet would be helical and their space would be filled with such vortices and their axes would all be parallel to a given direction. He assured readers that he didn't mean real rubber bands.

Sir J. J. Thomson

In 1893, J. J. Thomson[3] was saying that within the interior of a magnet there is a mechanism connected with the magnet which exerts a fan-like action, driving the negative tubes in one direction and the positive tubes in the opposite. Sir J.J. Thomson, who proved that the electron existed in 1898, suggested in 1907 that the winds of a LeSage's ether could form moving fields of force when it passed through moving electrons.

Without prior knowledge of Sir J.J. Thomson or LeSage, this author, in 1961, invented the LeSage ether on his own. He had read that Einstein had no ether and he believed that light waves demanded an ether.

This is exactly the idea that occurred to this author after he had decided that a pattern of helices would be impressed on the ether winds of a LeSage-like ether, when that wind passed through electron orbits. One key idea here would be that, when two to many orbits are lined up in a row, the resulting waves would be stronger in these two directions than in any direction where they were not aligned. J. J. Thomson had the concept of a fan-like mechanism which would drive the positive and negative tubes in their opposite directions. Is there much doubt that, had he known of the electrons in magnetic orbits, he would have ascribed to them that fan-like action? Waves turning in the same direction at the two ends of the magnet should have one common cause.

Thomson also said, "In a steady magnetic field where there is no free electricity, the Faraday tubes must be closed, exception being made , of course, of the short tubes which connect together the atoms in the molecules in the field. Since, in such a field, there is no electromagnetic intensity, there must pass through each unit area of the field, the same number of positive as negative tubes, that is, there must be as many tubes pointing in one direction as in the opposite direction.... We shall suppose that the positive tubes are moving with the velocity of light in one direction and the negative tubes with equal velocity in the opposite direction." At that point he did not mention what happens to the oppositely flowing tubes when they reach the ends of the magnet.

Leigh Page

In 1914, Leigh Page[16] published a theory in which an ether, moving in all directions, at the speed of light, would cause charged particles [electrons] to emit tubes of force which would radiate from the electrons in straight lines in all directions.

Whittaker in his history[17], used words which were not in Page's 1914 paper to describe Page's theory. "Page proposed to regard an electron, when viewed in an inertial system in which it was at rest, as being like a sphere whose surface is studded with emitters distributed uniformly over it; each emitter is continually projecting into the surrounding space a stream of corpuscles, each corpuscle moving radially in a straight line with the velocity of light; it is assumed that the emitters have no rotation relative to the inertial system. When the electron is in motion in any way, the stream of corpuscles that have been ejected from one emitter at successive instants, form a curve moving in space which, as Page showed, is a line of electric force in the field, due to the moving electron."

Page emphasized that the fields leaving the electron would be moving, not stationary. He also said that the total ether strain at any point will be the vector sum of all the strains due to all the particles whose fields extend to the point in question.

Ebenezer Cunningham

In the same year, 1914, Cunningham[18] said that the transfer of energy represented by the Pointing vector cannot be identified with the rate of work of the stress in the ether unless the ether is supposed to be in motion. He also stated that, "For a moving point charge [electron] the ether moves as if continually emitted from the charge with velocity C, every element travelling in a straight line after emission."

Einstein's Theory

The evidence for the validity of Einstein's relativity theory has seemed to be overwhelming. Eighty years of criticism has done little more than make pin pricks in the confidence of the great majority of the physicists. The number of proofs is large and removing one proof at a time does little to embarrass the theory. Part of the problem is that it is inherently difficult to argue about the things which are happening in another dimension. It is equally difficult to argue about the powers of those imaginary observers. Another problem is that so many of the critics offer no alternatives. The biggest problem, however, is the refusal of the journals to publish such criticism, even when it comes from qualified, working, physicists.

Paul Gerber

In 1898, Paul Gerber[21] modified Newton's laws of gravitation to include travel time for forces moving at the speed of light. Using the revised laws, Gerber was able to account for the discrepancy in the calculation of Mercury's perihelion. He did not need to tamper with our common notions of time and space. Ernst Mach[22] mentioned Gerber's paper in the 4th edition of his book on mechanics. In 1904 in the 5th edition he mentioned it again. In 1915, when Einstein accounted for the discrepancy in Mercury's perihelion, it was acclaimed to be proof of dilated time and curved space. In 1917, Ernst Mach[22] arranged for the publication of a complete copy of Gerber's paper in the same journal that published Einstein's 1905 paper. If he was trying to set the record straight, he failed. Books on relativity theory do not acknowledge Gerber's prior paper on this subject.

Werner Heisenberg

In 1926, Werner Heisenberg[23] had an interesting conversation with Einstein; he reported this in 1971. He explained to Einstein that electron orbits were banished from quantum theory because they were not observable. Einstein replied with the following two passages, as related by Heisenberg.

"But you don't seriously believe," Einstein protested, "that none but observable magnitudes must go into a physical theory?"

"Possibly I did use this kind of reasoning," Einstein admitted, "But it is nonsense all the same. Perhaps I could put it more diplomatically by saying that it may be heuristically useful to keep in mind what one has actually observed. But on principle, it is quite wrong to try founding a theory on observable magnitudes alone. In reality the very opposite happens. It is the theory which decides what we can observe. You must appreciate that observation is a very complicated process...."

Petr Beckman

In 1987, Petr Beckman[24] published a book with the title, "Einstein Plus Two." In this book, Beckman has given alternative explanations for twenty of the proofs of Einstein's theory. Beckman's proofs seem more logical than those of relativity theory because Beckman's do not need dilated time and warped space.

Beckman had also discovered that paper by Paul Gerber that dealt with the perihelion of Mercury. He laid out the mathematics of Gerber's paper. He also suggested an idea which this author agrees with very much. He thought that the successes of Einstein's theory were due to the manner in which the Lorentz transformation equations compensated for an inverse-square law that becomes inaccurate at high velocities.

Here it will be phrased differently. The Lorentz transformation equations, in a very roundabout way, were able to correct a flaw in Newton's laws. Newton assumed that the to-and-fro distances were identical. This would be true if travel time were instantaneous. Because of orbital movements and changing exposures to forces from many directions, the to-and-fro distances are almost never identical. The distance along a gravitational vector may be larger or smaller than the instantaneous distance. The actual pull of the sun's gravity depends on the inverse-square law as it affects the actual distance that the sun's force will travel.

David Bohm

In 1957, in David Bohm's[50] book, "Causality and Chance in Modern Physics," he criticized the usual interpretation of quantum theory, where Heisenberg, Bohr, and others insisted that there could be no need for sub-quantum causes or sub-quantum mechanisms Bohm claims that those physicists were far too hasty in their conclusions. Their arguments have no more basis than numerical discussions of angels dancing on the head of a pin. David Bohm also suggested that the quantum orbits could be arranged along pilot waves. This was taken care of by this author's united field equation in chapter l.

Richard Feynman, in his new quantum theory, for which he and two others received the Noble Prize, managed to get rid of the worst absurdities, those put in by Heisenberg, Bohr, Dirac, Einstein, and others. There was the statement that the electron was a point particle with infinite weight. There were no causes below the present level so there was nothing more to discover. There were measurements you could not make because of the uncertainty theory. Feynman disliked Dirac's use of complex numbers+ like little i and didn't use them himself. Feynman, however. used most of relativity theory.

The Kennedy-Thorndyke experiment of 1932, which had the same null results as the Michelson-Morley experiment despite the fact that in the K-T case, the two paths were widely different in length, while in the M-M case the lengths were the same. In the case of M-M the two contractions could be X and X^2 or it could be X in the forward direction. in the K-T case the contractions could have been X and Y^2. The K-T null results should logically prevent scientists from continuing to use M-M figures as a ratio, but that would destroy the foundation for relativity theory.

Chapter 8

Gravitation is a Product

Newton is quoted as saying that the gravitational attraction between any two particles of matter, varies directly with the product of their masses, and acts along the line that would join their centers, with a strength that varies inversely with the square of the distance between their centers. Many years ago this author found himself speculating on why gravitation varied as the product rather than as the sum of the two masses. It had occurred to him that the gravitational attraction could be between the ultimately smallest particles of matter.

If all gravitational matter is divided into uniformly small particles, we would have an explanation for why a lead ball falls at the same speed as a wooden ball. The particles that are falling are all of the same weight. If the smallest particles are all the same weight they would not need to be identical otherwise. In a vacuum, where the friction of the air is nonexistent, gluing particles together does not result in the particles falling faster. If two units here are attracted to three units there, we would have two groups but, if the unit particles are reacting, all with each other, we would have six pair of straight lines between their centers. Two pairs of lines would extend from each of the three and three pairs of lines would extend from each of the two. We can say that the number of pairs of gravitational lines varies directly with the product of the two numbers which represent the number of units in each group. We can ignore the lines which are internal to the structures of the two groups.

If we restructure the five particles so that there are four particles here and one particle there, we have the same sum, five, but the number of lines which would join each particle in one group with all the particles in the other would now be four Pair instead of six pair.

Others have thought along this line. Petr Beckmann[25] said that it is interesting to contemplate the product of masses, (charges), in the Newton-Coulomb law. Beckmann said that masses and charges are evidently not team players; they interact individually, each particle of one group with each particle of the other group. Rudolph T. Lee[27] said that the fact that two material bodies attract each other with a force which varies directly with the product of their masses is an indication that gravitation is an action between the ultimate particles of matter.

The above concept of ultimate particles is practically unknown. It is not mentioned in discussions of the mystery of why leaden balls and wooden balls fall at the same speed. It is also not mentioned in discussions of why inertial mass and gravitational mass are equivalent. The equivalence would be due to the situation in which the reacting units are one and the same. The momentum caused by gravitation is the same momentum which is caused by billiard balls striking each other, or that which is caused by electrical or magnetic forces. It would be the smallest particles which receive momentum, carry momentum, and deliver momentum. The capacity of matter to carry momentum, in measurable amounts, is what makes the laws for the conservation of energy, possible.

Pairs of Lines and Gravitational Formula

There is a complication in dealing with the number of straight lines between two gravitating bodies. The straight lines of Newton's formula are really pairs of lines. Each body will receive a line of force as well as send a line of force. If the two lines are always the same length, there would be little significance in their existence. If the two bodies have any common motion along the two lines that join them, a difference in their lengths will exist. This difference is accentuated by the inverse-square law. A small decrease in the distance traveled can make a larger increase in the pull exerted.

One way of making a comparison between gravitational force and electromagnetic force would be by comparing the gravitational force of the earth on our moon and the momentum which is created, for the most part, by the difference between the pulls of the electromagnetic bonds which create that momentum in the moon. If two bodies are moving in the same direction, one following the other, one factor which affects the pulls which they have on each other is the common speed which they share in that same direction. The effect of such a gravitational pull may be very minor compared to the previously acquired momentum which may reside in the electromagnetic bonds of those bodies.

When gravitational forces pass each other, going in opposite directions, they move away from each other at twice the speed of light. The ether winds which carry all the field forces, move through each other in all opposite directions at twice the speed of light. All bodies and all particles move relative to the opposite ether winds. If a movement is negative relative to one wind, that movement may be positive relative to the wind from the opposite direction.

Gravitational forces travel in straight lines in all directions at the speed of light. Electromagnetic waves spin around straight lines which travel in certain directions at the speed of light. Gravitational forces exert pulls on particles which happen to lie in their path of travel. Electromagnetic waves exert their influence on electrons, electrons in orbits, and on atoms which have a propensity for reacting to orbits of certain sizes. Gravitational forces seem to resemble very small bubbles of vacuum-like space which are carried by the ether winds in all directions. Electromagnetic waves seem to resemble very small bubbles of vacuum-like space arranged in spirals or other shapes, around their line of travel.

A new theory of gravitation should take into account the time needed for waves to travel in the two directions between two moving bodies. While it may be important to know which body is following and which body is leading, it may not be necessary to figure the pull from each body separately in each two-way relationship.

The following equation expresses the two-way gravitational force between two bodies when no other bodies are being considered. The gravitational attraction between them will increase, when their mutual speed increases. The following is a formula for calculating the force at each changing speed.

$1/2\ M_1\ M_2 / d^2 (1 - V/C) + 1/2\ M_1\ M_2 / d^2 (1 + V/C) = \text{Force.}$

$M_1\ M_2$ gives the total number of pairs of lines. 1/2 of the lines are the shorter $(1 - V/C)$ lines and 1/2 of the lines are the longer $(1 + V/C)$ lines. In order to add the two fractions together we must make the two divisors, identical. To do that we can multiply the first fraction by $(1 + V/C) / (1 + V/C)$ and mutiply the second fraction by $(1 - V/C) / (1 - V/C)$. The two divisors will contain the factors, $(1 - V/C)$ and $(1 + V/C)$. The product of the sum and difference of two numbers will be the square of the first less the square of the second, $(1 - V^2/C^2)$ in this case. In the dividend, the two V/C s will cancel out. As a result, the force will equal $M_1M_2 / d^2 (1 - V^2/C^2)$

The actual contraction, the subtracted amount, will be V^2/C^2. These figures will not hold when speeds get close to the speed of light. At high speeds it would be necessary to use a calculated table of values. If a real problem is to be set up, the velocity, the product of the two masses, and the distance, d, must be entered into the formula. The denominator would reduce to d^2 when $V = 0$. The revised gravitational law would mean that, with any increase in speed, the pull between the two bodies would increase and there would be a real contraction in the line of motion, a contraction which would increase as the speed, V, increases.

Before leaving gravitation, another point should be made. The gravitational force that holds the earth to the sun, that continuously pulls the earth out of its momentum in a straight line to cause it to orbit the sun, travels from the sun to the earth at the speed of light. Paul Gerber[21] proved this last statement with his solution to the problem of Mercury's perihelion.

This pull is not like that of the rope that you might use to twirl a weight around your head. Gravitation reaches out everywhere in space, not just to the gravitating bodies that are affected by it. Practically all of this pull merges with the multidirectional ether where it may eventually be neutralized by inequalities in other directions or spread to the disappearing point by the radiant ether. From this last you may surmise that it would be difficult to balance an energy equation that involves gravitational attraction.

Chapter 9

The Ether

Everything in this book started with a rather simple idea, a theory of an ether that could carry light waves. Light waves can pass through each other in all directions, therefore the ether should have the same power. Naturally the ether winds that the light waves wave in should also move at the speed of light. The light waves travel from the most distant parts of the universe. The small parts of the waves, the waves must have parts, must be very small. Particles which can blow through each other in all directions can have mean free paths with lengths which expand as the postulated diameter becomes smaller. According to the gas laws if the diameter of the particles is 1×10^{-50} cm, the mean free path can be 1×10^{150} cm long, which would be the inverse of the cube of that diameter.

The discovery, by this author, that waves could be formed in such an ether, was followed by a search of the history of theories of ethers. A history was discovered of a 200 years old theory of a multidirectional-energetic ether. LeSage, a Swiss physicist, in 1744, proposed an ether theory that was still being discussed a century and a half later. George LeSage hypothesized that very small ether particles, moved in straight lines, in every direction, at enormous speeds. The particles of matter would be pressured from every direction by the much smaller ether particles. If two particles of matter partially shielded each other from the constant bombardment, each would have less pressure on the side nearest the other. They would be pushed toward each other.

In 1873, in an article in the Encyclopaedia Britannica, Clerk Maxwell[28] criticized LeSage's energetic multidirectional ether theory by saying that the action of such an ether would raise the temperature of the universe to a white heat and raise all bodies to enormous temperatures. He was saying that this heat would be due to the pressure caused by the enormous bombardment.

Maxwell thought that this ether would cause heat, but there are other ways to look at the problem. LeSage's ether would produce motion and only the motion would produce heat. The vibrations of heat within molecules and atoms could arise only if the parts moved relative to each other. There could be no vibrations and no heat waves without this relative motion. The ultimately smallest particles of matter would be subjected to the pressure of the LeSage ether. If these smallest particles have no parts to vibrate and no apparatus to create heat waves, how could we be aware of any heat caused by the pressure on them? How could pressure be made on particles of mass by the pull of gravitation? How does a pull become a push?

The ether particles are a very unknown quantity. How was Maxwell so sure that the ether particles would cause heat before they caused motion. There is the possibility that the ether particles are tiny magnets. There may be other means by which the ether's energy may be transferred to mass.

Einstein believed that photons are strings of particles. They seem to able to string themselves together, as if along a strong thread of magnetic force. There must be some way in which the individual units of energy are tied together as they swing orbitting electrons loosely around elliptic orbits. The energy of an electron orbit lies around that orbit like a string of pearls. They also carry themselves across the universe with their various and exact spectral charges. A photon could be called a string of Plank's pearls.

Maxwell's 1873 criticism was very widely read because it was reprinted in other editions, up to and including the 1890 edition of the Encyclopaedia Britannica. Maxwell, however, did not kill the LeSage ether theory. Aronson[29] tells of S. T. Preston and Sir William Thomson working hard to revive the LeSage theory in the 1880s. S. Tolver Preston[11] published two editions of a full length book describing the many advantages of the LeSage's theory in explaining gravitation. Small changes could meet lots of arguments. Whittaker[18], in his history, told of Sir J. J. Thompson in 1907, Ebenezer Cunningham in 1914 and 1915, and Leigh Page in 1914, using the LeSage ether to explain the lines of force in energy fields. They were theorizing that electrons could create waves in the ether winds which would form fields of force. The LeSage theory was finally put to rest when Einstein's theory took its big boost at the time of the eclipse of-the-sun expedition.

Every point in space would be a radiant center for the ether which moves in every direction. A group of ether units or particles which happen to be blowing in the same direction will be called an ether wind. An ether wind which is moving perpendicularly through the plane of an electron's orbit has a special relationship to that orbit. There will be two perpendicular winds blowing through the plane of each orbit. They will be blowing in opposite directions. When the ether winds blow through an electron orbit the electron will create helical paths or helical waves in the two winds. The winds are continuous, so long series of waves can be created.

When two orbital planes are adjacent to each other, the south wind of the first will enter the north end of the second, and the north wind of the second will enter and blow through the south end of the first. The orbit of an electron will turn at the same speed as the turn of the helical wave. If two orbital planes are one wavelength apart, the end of the first helical wavelength can intersect the orbit of the second electron at the exact spot where the second electron started its first orbit.

The electron travels around the atom in its orbit in the same amount of time that the wind needs to carry the wave from one orbital plane to another. The electrons move at a slower speed than the waves, therefore the electron's path must be much shorter. As the electron moves around its orbit the wind picks up a unit of pull as each diameter of the electron crosses the path of the wind. The pull of the electron on another electron has been likened to a shadow or the pull of a vacuum. If an electron and a wave cross each other at a right angle, the wave will exercise one unit of pull on the electron. If a wave follows an electron completely around its orbit the waves can exercise thousands of diameters or units of pull on the electron. At the same time the wave will be picking up additional units of pull that can be exercised if additional orbital planes are encountered.

Magnetic bonds such as have been described above, can be between two atoms or along thousands of atoms. In Chapter 9, phase control and the contraction mechanism for magnetic waves are described.

The radiant nature of the ether causes the inverse square weakening of the directional electromagnetic waves. 4 pi r^2 is the formula for the expanding surface of spherical radiation.

The radiant forces which control the orbits are divided along the length of a 2 pi circumference and weaken inversely with the square of the distance to a point on that circumference, In atomic research h is divided by 2 pi. That can be a division of the energy in one orbit. That could give us the angular velocity for every point on the length of that orbit. The inverse square law developed very naturally from the inverse r^2 above. Electromagnetic waves are directional and result from magnetic orbits and electrical spins. Gravitational waves are not directional, they are radiant.

This ether has many useful features. The square of the speed of light times the mass of a small ether particle allow the development of a measurable momentum. The radiant nature of the ether explains all kinds of radiation. The very small size of the ether particles allows them to discover the smallest interstices of electrons or other matter and there deliver momentum. The small size of the ether particles explains the ability of remnants of light waves to travel billions of light years to end up disturbing an electron on the surface of a photographic plate.

There were many ether theories in the nineteenth century. For the most part they theorized about stationary ethers or jelly-like ethers. The Michelson-Morley experiment was supposed to reveal the earth's motion through a stationary ether. When that experiment failed to show that movement, the general reaction was that all ethers were ruled out. That was a measure of how little favor the energetic LeSage ether had at that time. The Fitzgerald-Lorentz contraction hypothesis, which seemed to save the stationary ether, was criticized for being too flimsy, too "ad hoc". Faith in ether theories lost favor. When Albert Einstein suggested that we forget all about ether theories, his suggestion was taken as a command by most of his followers.

Generations of physicists, and the journals they publish in, have avoided that subject. There is great loyalty to Einstein's relativity theory. It seems that only a historian will mention the usefulness of ether theories. Einstein did seem to relent later and said that an ether theory might be necessary, but by that time the quantum theory had taken over with its ban on the possibility of the existence any kind of reality or causes or structures below our present knowledge.

The biggest mysteries of the universe are left unexplained. The physicists ignore these problems as if they were already solved. There seem to be spoken and unspoken hypotheses that some things just happen to be in the nature of things and therefore need no explanation. Of all the mysteries that an ether theory should deal with, the biggest one would be the how and where of the immense store of energy that seems so readily available for all the phenomena involving motion and heat.

In earlier times, physicists found many clues on which to base ether theories. In free space, light waves move in straight lines. The gravitational force between two bodies is measured along the straight line that would connect their centers. Momentum moves mass in straight lines until that mass meets a force that can change that course. George LeSage[10] had advanced an ether theory that incorporated these straight lines.

The main evidence for the radiant nature of the ether lies in our radiant universe. This radiant nature is expressed in many ways. Gravitation furnishes one illustration. The gravitational attraction of every single particle seems to reach and influence every other particle in its vicinity, no matter in what direction. The manner in which the earth's gravitational field seems to exist at every spot in the path of the moon, is part of the evidence. The gravitational radiation emitted at any single instant must be spherical or it would not have the ability to extend in every possible direction at the speed of light.

Light is another example of a natural force moving in every possible direction. Light from billions of stars can converge on the small eyepiece of a telescope. Every speck of free space seems to be a radiant center where light flows in from all directions and out in all directions. Light from a galaxy a billion light years away will pass through billions of such radiant centers every second of its journey.

One might ask how such a thing could be, and a possible answer is as follows. The light photons are composed of groups of very small particles, or they are composed of groups of holes in a moving, radiant ether. They have to be very small to have such a long free path. This refers to the average distance that a particle could travel without hitting another of the moving particles. If the size of the ether particle is $1/X$ the length of the mean free path will be X^3.

The mathematics would be somewhat as follows. A one inch diameter ball in one inch square box has the freedom to stay put. In a two inch square box it has eight times as much room. In a three inch square box it has twenty-seven times as much room in which to travel. The effect would be similar if the balls became smaller and the box retained its size. The LeSage ether theory, as proposed by S. T. Preston[31] would obey the gas laws with the entire universe as the container.

The electron's field is also radiant. Its magnetic field extends in every direction. The well-known iron filings test shows that a magnetic field completely surrounds the magnet. We know that the electron radiates to the gravitational field because the electron has weight. That the electron radiates in all directions is accepted by scientists[17].

Endlessly and forever every particle of mass continues to radiate that field force called gravitation. Endlessly and forever protons and electrons can pull on each other. Why can't the physicists see that it is the orbital momentum carried by the ether that stops electrons from falling into the nucleus.

Einstein, at the time he invented his fourth dimension and abolished the relativity of the speed of light, delivered a third blow to common sense by effectively abolishing the ether. He had no need for an ether theory, and he said so.

In the nineteenth century physicists thought that the first step toward understanding the field forces would consist of forming a theory of an ether. Magnetic and electric forces gave great evidence of being waves in an ether. The nineteenth century theorists thought that they needed an all-pervasive ether in which those waves could wave.

This author, by disregarding the prohibitions of relativity theory and by following the laws of algebra and the laws of common sense, and by following a unified field theory which he would prefer to call a paradigm, has discovered solutions or explanations for a tremendous number of experimental. results, which could also be called facts or numbers.

Chapter 10

Clues To Magnetic Structure

The poles of the magnets furnish most of the clues about the inner structure of magnets. When a magnet is cut in two, the freshly cut ends acquire poles in a way that gives each piece both a north and a south pole. This can happen, even if the pieces become very small, because the magnetic force is produced by long rows of atomic magnets with their north poles facing south poles. Small magnets can be one atom long. Modern science credits an electron in a magnetic orbit as being similar to a current moving around in a electromagnetic winding. Very strong magnets are made by winding coils of wire around soft iron cores and sending a current around the coils. The current consists of billions of moving electrons.

In 1848 Faraday[32] discovered a strong connection between magnetic lines of force and crystalline bonds. He showed that when a crystal of bismuth is placed in a field of uniform magnetic force, it sets itself so as to have one of its crystalline axes directed along the magnetic line of force.

Magnets can set themselves along the magnetic line of force of other magnets. A diamagnetic crystal seems to have a similar capacity inasmuch as it can react to a magnetic line of force. A bar magnet has long rows of atoms with electrons in magnetic orbits. A crystal has long rows of atoms in lattice like patterns. One big difference is that, in the bar magnet, the waves become strong enough to emerge from the magnet. The atoms in a diamagnetic crystal do not create such strong waves.

It is not claimed that all bonds are created by electrons in magnetic orbits. The bonds that bind electrons to atomic nuclei are waves created by the spinning of electrons, interacting with waves created by spins within the nuclei. These are electrical in nature and should be of much smaller wavelengths than the magnetic waves which are created by the gross orbits around the atom.

It is also possible for atoms to be bound together by electrical bonds. In order for this theory of double helical bonds to be thoroughly consistent, the electrical bonds should be helical also, but on a much smaller scale. At the very least, the electrical bonds should transfer momentum from the ether to the mass of the atomic parts.

One of the early clues to the nature of magnetic structure was the discovery of the rotary nature of the magnetic field. In 1845, Michael Faraday[33] discovered that when a magnetic wave was sent through a slab of glass parallel to a beam of polarized light, the light beam was rotated due to the presence of the magnetic field. It was discovered that the north pole of a magnet rotated the beam in one direction and the south pole of a magnet rotated the beam in the other direction. The amount of rotation was proportional to the strength of the magnet. When the magnet was removed there would be no rotation.

Another clue was discovered when groups of electrons were discharged from a hot electrode to fly through a narrow opening to hit a screen or photographic plate. When their paths were made to pass along a magnetic field their point of impact was rotated to the right or left, depending on which magnetic pole supplied the field.

Several conclusions can be drawn from these results. The first one would be that there is some mechanism inside of the magnet that can cause the magnetic field to rotate. The second one would be that magnetic waves can cause electrons and atoms to move. This seems to prove that the magnetic field can carry momentum and impart it to electrons and atoms. This evidence can be added to the evidence in Chapter 2 for momentum at less than the speed of light. Another conclusion would be that turning a magnet around causes the apparent rotation to change from clockwise to anticlockwise. This has been interpreted by some persons to mean that the fields from the two poles of the magnet are turning in opposite directions. Of course they turn like the two ends of an axle. The handedness of the rotation depends on the position of the observer.

D. S. Parasnis[34] gives a very puzzling explanation of the lines of force which a bar magnet displays in the iron filings experiment. The following is transposed to some extent. "At first glance it appears that force originates on the two poles of the magnet. The force seems to travel outside the magnet, in a curved line, from the north pole to the south pole. In reality, they continue within the magnet and form closed loops. The forces within the magnet can be measured by cutting narrow slots at various spots and inserting small dipoles. The force within a long magnet can be very uniform along its length except immediately at the pole faces. The measured strength just outside and inside the poles is one-half of what it is at the center of the magnet."

He explains the doubling of the strength inside the magnet by saying that a demagnetizing field is created at each pole and this field extends to the center of the magnet. We could suppose that he meant that the demagnetizing field from the other end of the magnet also extends to the center of the magnet. He doesn't mention this. He does say that the demagnetizing force turns to zero at the center of the magnet.

He says that the net field just inside each pole is one half of what it is at the center. He says, "If we consider the entire picture we find that the lines of the magnetizing force do not form closed loops, but have a discontinuity in their direction at the end faces of the magnets." He may have meant that the demagnetizing fields start there. He says that demagnetizing forces arise whenever the lines of a magnetic field are required to cross an interface between two magnetic media.

Parasnis has a statement about magnetic attraction and repulsion which seems to foretell the present theory. He says, "The attraction and repulsion between two magnets is due to the non-homogeneity of their fields. The force exerted by a magnet on the nearer pole of the other magnet is numerically greater than that on the farther pole." In this theory the non-homogeneity is due to the simple fact that turning one of two magnets end to end reverses its spin. We can visualize then that the field which leaves each magnet to invade the other magnet is spinning in the opposite direction to the spin of the interior field of the invaded magnet.

In this theory, the double helix in the magnet is composed of two rotating fields which pull in opposite directions. The pulls are in perfect balance with each other, that is, any imbalance is in accordance with the relative motion of the system. The invasion of a field which is rotating in the opposite direction kills or weakens the interior field which is pulling in the near direction. The field which would pull the magnet in the opposite direction is not weakened. The balance is destroyed. There is no repulsive force. Just like in the case of gravitation, every force is a pull.

A 1986 book, whose title might lead you to think that it was very inclusive, Handbook of Magnetic Phenomena by Harry E. Burke, does not include the term, demagnetizing, in its index. This just points out that many of the facts mentioned in this book are not likely to be available to modern students. To the physicists, the above may be included in the innate properties of magnets and they would think that it would need no discussion.

The demagnetizing field is mentioned in the National institute of Physics handbook of 1957. It is stated that the ends of a magnetic rod cause magnetic fields to appear is all parts of the rod[35]. The term is just an expression that has been invented to describe a very mysterious phenomena. They seem to have no idea why it occurs. Possibly the reason for the lack of theorizing along this line can be attributed to the illusion that the field of a bar magnet curves around the sides of the magnet to enter the other end. Research in the literature has failed to turn up any reference to the possibility that the demagnetizing effect could be due to the existence of two magnetic waves within the magnet with each starting at one end of the magnet and emerging from the other end. It is, however, quite probable that somewhere, someone, has voiced such an idea. If they submitted the idea for publication, and it was rejected, we would never know about it.

The concept that electrons build magnetic fields by emitting elements of force which move at the speed of light in all directions is centrally important to the mechanisms suggested in this book. This is an idea that was first proposed by Leigh Page in 1914, (See page 31.) Of course we could do as the physicists do; we could just take it for granted that the electrons emit these forces. It is not necessary to postulate an ether. The mathematics of the double helix would not need any changes. The physicists prefer mathematical descriptions, devoid of mechanical explanations. If they ever stopped to think about the radiant nature of electrons they would probable call it another innate property of matter.

The physicists do not seem to realize that all the separate assumptions are separate postulates. Even the statement that there are no causes is a postulate. Yet they claim that postulating an ether is a very serious step. It may be that they think it a matter of deep impertinence when someone claims that thousands of physicists could possibly be wrong; there might be an ether.

If you prefer an understandable mechanical universe instead of one composed of mathematical symbols, consider the multidirectional and energetic ether which LeSage proposed in 1744. His ether provides a mechanical explanation for the production of all kinds of radiation. A moving electron will radiate its force, in all directions, at the speed of light. The electron will have no further influence on the course of the radiation which it emits. The radiant ether carries the elements of the electron force in straight lines in all directions.

The ether winds can explain the magnetic waves. An ether wind could be defined as a set of ether forces which are moving, as a group, along parallel lines, in the same direction. A moving electron, as it cuts across the parallel lines of an ether wind, will emit forces which will move along those lines. The emissions, at successive instants, will form moving lines of force. The movement of the electron as it curves and turns will cause curves and turns in the lines of force which develop in an ether wind. These curves and turns can keep their shapes and proportions as they move away at the speed of light. Due to there being parallel winds blowing in every direction, there will be little significance to such a wave unless the curves and turns of a line of force should follow an electron in its path long enough to influence it, or if it should join a number of similarly curved waves lying in the path of an electron. Lines of force which move from electrons in all directions, and this includes all opposite directions, will cancel each other unless some coincidence of path occurs. This is why electron orbits selectively absorb certain light waves and not others. The electron movements which create certain size waves can also have the ability to absorb energy from waves of that size.

Iron filings have been used to test the nature of the fields surrounding a bar magnet. Iron filings are spread evenly over a piece of paper and the paper is placed over a bar magnet. The force of the magnetic field arranges the slivers of iron as if they were magnetic needles. If you study the pattern created by the magnetic field it will appear that the field flows out of the magnet at one end, curves around to flow along the side of the magnet and enters the other end of the magnet. This is the way that it appears. The actuality can be that the elements of the field move out in straight lines in all directions. They are carried by the radiant ether.

The magnetic fields are affected differently as ether winds move in different directions across the orbits which create the fields. Ether winds, moving perpendicularly, in opposite directions, through the planes of magnetic orbits, will have helices imprinted on them. Winds moving in different directions will form waves with different curves. A wind moving edgewise across magnetic orbits can have sine curves imprinted on it. The waves emitted in the lateral direction are organized also and this organization can be impressed upon the iron filings which lie in that direction. The patterns in the field can control the directions of the orbits of atoms in the iron filings. It has been well proven that magnetic fields can control the movement of electrons. It takes no jump of the imagination to visualize the magnetic fields controlling electron orbits. Pieces of iron which are magnetized, line themselves up, each north pole seeking a south pole. The fact that the slivers of iron seem to follow what seems to be a line of magnetic force, only proves that they are magnetized.

When two south poles or two north poles of bar magnets are set to face each other, they seem to repel each other. The real explanation can be that the magnets pull themselves away from each other. The only force they have is a pull. The near end of each magnet is invaded by a wave from the other magnet which is spinning in a direction opposite to the direction in which the electrons are orbitting in the invaded magnet. The invading wave would reduce the pull on the electrons and upset the normal balance between the opposite waves. The magnetic waves which are pulling in the opposite direction will now pull the magnet away. The waves pull and guide the electrons around their orbits as well as pull perpendicularly on their orbits.

When two bar magnets are arranged so that a north pole faces a south pole, the external fields of the two magnets pull the magnets together. The invading waves, as they move into the other magnet, will find electron orbits spinning in the same direction. The pulls of the invading waves are added to the existing pulls and the balance is overwhelmed in the direction of closing the gap between the two magnets.

Weak magnetic waves can become strong bonding waves. The forces will be much stronger when the poles are in close contact. The inverse-square law operates in this case. If the distance apart drops from one centimeter to a millionth of a centimeter the bonding force would become very much stronger. At present, relativity theory does not allow the inverse-square law to operate at the level of the electron bonds. This book presents many pieces of evidence that prove beyond a reasonable doubt that the inverse square law does apply to magnetic bonds.

Chapter 11

Transformation Equations

The many successes of relativity mathematics can be credited to one simple fact. The Lorentz transformation equations, which form the foundation for Einstein's theory, plus a large body of complicated relativity mathematics, provide an equivalence for an error in Newton's laws of gravitation. Newton's laws were not fully accurate because they failed to take into consideration the time needed for waves or forces to travel in the two directions between two moving bodies.

Take the case of two heavenly bodies moving through space and reacting to each other's moving presence. Each body emits waves which travel at the speed of light. These waves are attractive forces which lose strength inversely with the square of the distance traveled. At twice the distance they will be one fourth as strong. If a wave is meeting a body which is coming toward it, the distance which a wave travels will be shorter. If a body is moving away, so that a wave must chase it, the distance which a wave travels will be longer. In algebraic terms, the speed of the waves can be labeled, C, and the speed of the bodies can be labeled, V. If one body is following another, both bodies moving at the speed, V, the wave that moves to the rear will meet the following body by travelling the distance, C – V. The wave from the rear will overtake the leading body by travelling the distance, C + V. If the speed of light, C, is given the value, 1, the speed of the body, V, can be written as V/C, a fraction of the speed of light. (C – V) will be (1 – V/C) and (C + V) will be (1 + V/C).

Einstein effectively abolished (1 + V/C) and (1 – V/C) when he postulated that the speed of light is constant. It is forbidden to add or subtract V to or from the speed of any wave force that moves at the speed of light, C. No matter how fast an observer or any material object is moving, the speed of light will remain constant, not relative. Einstein had already provided the correction for Newton's formula when he used the Lorentz transformations, so it was necessary for Einstein to abolish (1 – V/C) and (1 + V/C).

The importance of the transformation equations points directly to the importance of the Michelson-Morley experiment. There have been many proponents of relativity theory who have claimed that Einstein had never heard of that experiment when he wrote his 1905 paper. They seemed to think that discussions of the Michelson-Morley experiment were irrelevant to a discussion of relativity theory.

There are also many who have credited that experiment with being the basis for the origin of Einstein's theory of relativity. Einstein stated many times that he had based his theory on the transformation equations that Lorentz had proposed. Lorentz created those equations in direct response to the problems raised by the results of the Michelson-Morley experiment. Lorentz was trying to save the concept of a stationary ether which was discredited by the null results of that experiments[36].

Einstein's biographer, R. W. Clarke[37], reported that, in 1952, Einstein admitted that he must have been conscious of that experiment through the reading of the papers of Lorentz. However that may have been, it is important that the transformation equations were the direct result of the Michelson-Morley experiment and had a large influence on Einstein's theory, both in its origin and in its acceptance.

The factor $(1 - V^2/C^2)^{1/2}$ was used twice in the Lorentz transformation equations. These factors are in the foundation of relativity theory. No matter how complicated Einstein's equations become, you must remember that those transformation equations lie everywhere in the ancestry of relativity mathematics. The source of those equations, $(1 - V/C)$ and $(1 + V/C)$, has been disowned. They are the black sheep in the family tree. Expressions of the relative speed of light are forbidden.

The mathematics of $(1 - V/C)$ and $(1 + V/C)$, in their many forms, seem to be the only mathematics in this book. It is the nature of the subject. All bonds reciprocally function in two directions at once. This is what the field of physics seems to deny. The freedom to use this mathematics is the advantage this author has in not being a physicist. There is also the freedom of working without the pressure of time; the field, seemingly, is devoid of competition.

The mystery of the results of the Michelson-Morley experiment set the climate for Einstein's mysterious and logic defying proposals and was probably a major factor leading to their early acceptance. For most persons, the null results of that experiment were the only evidence they had for the necessity for abolishing the relative speed of light.

As a further proof of the importance of the Michelson-Morley experiment we quote again from Einstein's biographer, Ronald W. Clarke[38]. In 1920, at a dinner where Einstein and Michelson were guests of honor, Einstein addressed the following comment to Michelson. "You uncovered an insidious defect in the ether theory of light, as it then existed, and stimulated the ideas of H. A. Lorentz and Fitzgerald, out of which the Special Theory of Relativity developed. These in turn led the way to the General Theory of Relativity and to the theory of gravitation. Without your work this theory would today be scarcely more than an interesting speculation."

As an illustration of the popularity of the opinion that the Michelson-Morley experiment was the origin of relativity, we can quote from some of the essays on the special theory of relativity which were submitted in the, "Scientific American," essay contest[39] of 1922. G. W. Hemens said that the whole theory was based on the experiment by Michelson and Morley. After discussing the Michelson-Morley experiment, H. N. Russell stated that it was upon these facts that Einstein based his special theory of relativity. M. Francis says that, as a direct consequence of the Michelson-Morley experiment, he could state that light presents the same velocity to all observers, whatever their velocity of relative motion. H. T. Davis said that the Michelson-Morley experiment was the great physical fact upon which the theory of relativity rests. These four statements were in a book of fifteen essays selected from a total of 300 essays submitted.

Forty years did not seem to lessen the popularity of that opinion. The 1960 Encyclopaedia Brittanica[40] says that the failure to detect any effect of the motion of the earth on the velocity of light was the starting point of relativity theory. It is easy to find books on the special theory that spend two or more pages on the Michelson-Morley experiment. It is usually the only evidence which is the given for the necessity of a radical departure from classical physics. There is very little discussion as to why the Fitzgerald-Lorentz contraction hypothesis had to be replaced by an even bigger mystery, the dilation of time. They just said that there was an insoluble mystery.

Using hindsight, we can say that Hendrik Lorentz made a serious error when he created his contraction hypothesis and this error may have changed the course of history. This error showed up in 1932 when the Kennedy-Thorndyke experiment was performed. The error which Lorentz made was in using the ratio of the two contractions and using that as a contraction in the line of motion. Silberstein[36] reported in 1914, that Lorentz hesitated before using the ratio of the two contractions. He finally decided to use the ratio because it agreed with a theory in electrodynamics which he had published. The mathematics of the Michelson-Morley experiment indicated a possible contraction of their experimental apparatus in the line of motion and another contraction in the direction lateral to that line. Using the ratio of the two contractions as one contraction on the x axis, and adding a contraction on an imaginary time axis, Lorentz created his transformation equations. These were the equations which Albert Einstein used as the foundation for his relativity theory.

The failure of the Michelson-Morley experiment to reveal the motion of the earth through a stationary ether, was a major puzzle for scientists in the 1880's and the 1890's. Hendrik Lorentz and George Fitzgerald happened upon the contraction theory at about the same time. Fitzgerald, however, was never associated with the transformation equations. As it was first proposed, the ratio of the two contractions was used. Lorentz carried this arrangement over into his transformation equations. His choice also carried over into relativity theory.

The Michelson-Morley-type experiments were repeated many times and over many decades. To start a day's experiments they make a fine adjustment which causes the beams of light to return in perfect phase[41]. They then turn the entire apparatus in many directions in order to see if the light beams still return in phase. They always do. The mathematics of this will be gone over later. They had thought that this experiment would reveal the speed and direction of the travel of the earth through space. The amount which a beam would be out of phase was going to reveal the speed of the earth, and the direction in which the apparatus was pointed when that movement registered a maximum, would give the direction of the earth's movement.

In 1932, the Kennedy-Thorndyke[41] experiment, using arms which were widely different in length, had results which were still null. It made no difference which direction the apparatus was turned, the light beams always returned in phase. In this case they knew that the arms were not equal. It is obvious that a ratio of the two contractions would not explain the Kennedy-Thorndyke experiment. If we solve for the ratio of A^2/A, the answer is A. If we solve for the ratio of A^2/B, the answer would not be A unless A is exactly equal to B.

We must have a theory that explains both the Kennedy-Thorndyke experiment and the Michelson-Morley experiment. A two contraction theory would not depend on the two arms being of equal length and could explain both experiments. That theory of one contraction in the line of motion, which Lorentz used, was very much, "ad hoc." It provided an explanation for one case only. It didn't explain the results of the Kennedy-Thorndyke experiment. The Kennedy-Thorndyke experiment strikes at the very foundations of Einstein's relativity theory by casting strong doubts on the one contraction theory which Lorentz used.

The person who insists on a two contraction explanation had some responsibility of explaining how the two contractions can occur. That was the original purpose of the theory in this book.

The contraction in the direction lateral to the line of motion has a fairly simple explanation. This was solved when it was realized that the bonds between atoms did not shrink. The explanation involves the time needed for bonds to travel. That is a concept which is sometimes called, "delayed action at a distance." The mathematics will be covered later.

The contraction in the line of motion was not easy. The equations were there all along; it was the drawings that took the years waiting for an inspiration. However, the time was not all wasted. A reading of the history was going on and a lot of other items fell into place. The contraction mechanism is a beautiful example of synchronized movements, but there is no way to explain it in a few paragraphs. A whole chapter and three scaled drawings will come later.

Where would the special theory be if the Kennedy-Thorndyke experiment had been performed in 1885 instead of in 1932? Lorentz certainly would not have used a ratio for his contraction theory. Leigh Page might have had the inspiration to build the theory on which Sir J. J. Thomson had been working. Page had the key idea on the formation of waves. The whole course of history would be changed, The history of physics is very silent about research that probes the possibility that the Michelson-Morley apparatus contracted in the lateral direction as well as in the line of motion. It seems that there is a non-relativistic solution to the mystery of the null results of the Michelson-Morley experiment that has never been investigated, up to now.

Writers on relativity theory have said that every avenue had been tried and they were forced to turn to relativity theory. If there have been serious attempts to develop a two contraction theory, why have the historians not reported them? Kennedy and Thorndyke, in the introduction to the report on their 1932 experiment, mentioned that two contractions were possible, but when they summed up their results, they said that they had discovered another confirmation of relativity theory. In 1937, H. Ives[42] pointed out that a two contraction theory was still possible. Kennedy and Thorndyke had failed to notice or had neglected to mention that fact.

Chapter 12

Cartesian Coordinates

This chapter has been written to answer a certain question which comes from the field of relativity. If relativity theory is wrong, how come it has produced so many right answers? The answers are several. The Lorentz transformation equations contained the ingredients for the correction of the Newton's inverse square law. There were no laws for the use of imaginary dimensions and observers and Einstein had almost unlimited flexibility in the use of those factors. He realized the importance of those transformation factors and he used his imagination and mathematical skills until he achieved results that made sense. A discussion of the transformation equations and what they transformed will be undertaken here.

The x axis, the y axis, and the z axis are at right angles to each other. These axes describe the edges of a boxlike volume of space. There is one zero point for the three axes and it is called the point of origin. These axes are called the Cartesian coordinates. They can be used to designate the location and distance of a certain point in space. This will be a point that can be reached by traveling a given distance along x, making a 90° turn, traveling a given distance along a y axis, making a 90° turn, and traveling a given distance along a z axis. The spot reached by the use of these measurements is called the x, y, z point. The axes or edges come in pairs, as would be true in an ordinary box.

In using Cartesian coordinates, or axes, the math is simplified when all three distances are measured as if from the point of origin. The point of origin is then at one corner of a boxlike volume of space. The three edges which radiate from the corner of origin would be the x and y and z coordinates and the distance which we wish to measure, extends along the longest diagonal, to the farthest corner, of that same boxlike space, to the x,y,z point.

This was a method of locating positions and distances in the three dimensional space before Einstein changed the rules by adding a time dimension. In order to make the problem as simple as possible we will consider the simplest possible space. We will have a unit cube, a boxlike space, with three dimensions, each having the value, 1. Distances within this cube can be solved by Pythagorean equations. This unit cube will be used in the discussions that follow.

If we are considering only one dimension, we will have a straight line, and no formula is needed because, taking the square root of the square of a dimension, will return us to the original value. If we have two dimensions, we will use the formula for solving for the value of a diagonal; the square of the hypotenuse is equal to the sum of the squares of the other two sides. If we have a three dimensional space, the sum of the squares of the three dimensions will equal the square of the length of the longest diagonal within the cube. There is no formula for solving for a diagonal in four dimensions, one at a right angle to the other three. All the right angle dimensions have already been used.

To discover the length of the longest diagonal in this unit box we use a formula like the following. $(1^2 + 1^2 + 1^2)^{1/2}$, $= (1 + 1 + 1)^{1/2}$ and that is equal to the square root of 3. That will be 1.7320508, the distance to the point, x, y, z. This would be the answer if the system were sitting motionless in space.

For a three dimensional common-sense theory in which the coordinate system is moving at speed, V/C, the contraction along the X axis, the axis in the line of motion, would be $1 - V^2/C^2$, the contraction along the Y axis would be $(1 - V^2/C^2)^{1/2}$, or the contraction along the Z axis would be $(1 - V^2/C^2)^{1/2}$. The mathematics for these contractions is fully outlined in this book. These are the contractions which would explain the null results of the Michelson-Morley experiment. This is not a Newtonian or a classical reality; it is not the method of relativity; it is a three dimensional approach to reality.

Relativity mathematics does not add the squares of all four coordinates to find the square of the diagonal distance to a x, y, z, t point. There are problems if that takes place. There is an inconsistency in that the value of the fourth coordinate would be either l or 0 when the value of the motion, V, equals 0. It is obvious that a multiplication times 0 would give 0.

There is another problem. We cannot give the time axis the value of 1. The Pythagorean equation for a system of coordinates at rest would be, $(1^2 + 1^2 + 1^2)^{1/2}$ equals the distance to the x, y, z point. That computes to the square root of three. If the time axis has a resting value of 1, and if that is added to the other three coordinates before squaring them, the distance to point x, y, z, t, would have a rest-system value of the square root of 4. This would mean that adding a fourth dimension would be adding to the distance rather than causing a contraction. If the resting value of the t dimension is 0, that would mean that you could factor the other dimensions but you could only add to a zero.

In relativity theory mathematics, the time dimension is not added to the other three dimensions by a plus sign or subtracted by a minus sign. You might interpret it as a plus-minus sign or a sign of some sort of a reversal. It is the square root of –1 and it is quite often designated by a little i. New mathematical uses have been invented, as needed, in order to pass over the inadequacies of their abnormal mathematics. These are seemingly justified by their success in explaining experimental results. They are not justified until the two contraction explanation for the null results of the Michelson-Morley experiment has been thoroughly studied. In this book only normal mathematics is used and it is explained perfectly.

i is sometimes called an imaginary number but the way it is used puts it in the category of an operator. Operators include such symbols as x, /, –, +, (,), and many others. See Guy Murchie[43], in his book, "Music of the Spheres", pages 472-476,

In the course of that intricate mathematics this alteration may have occurred in the rules for coordinate systems. The time contraction may have been used to put an adjustment to the length of three regular dimensions before squaring and adding them. When the contraction on the time dimension is A and the contraction on the x dimension is A, the multiplication of the three coordinates by the time contraction would look like this.

$$A(x = A, y = 1, z = 1) = (x = A^2, y = A, z. = A).$$

The x axis is in the line of motion and the contraction in the line of motion is $1 - V^2/C^2$. The lateral axis is either y or z and the lateral contraction is $(1 - V^2/C^2)^{1/2}$. There are only two paths to contract in the Michelson-Morley experiment.

Chapter 13

Atomic Bonds

The fact that crystals resist crushing or stretching means that the bonds within a crystal are rigidly concerned with maintaining set distances apart. In crystals, there can be long rows of very evenly spaced atoms. That is very suggestive of spacing along the nodes of a standing wave. We know that adding energy in the form of heat can cause crystals to expand. That seems to indicate that the lengths of the bonds are controlled by quantum conditions within the atoms. Orbital levels or energy levels are related to quantum conditions, wavelengths are related to orbital levels and the rigid bond can be related to the wavelength. It is also to be noted that enough heat can destroy the bonds altogether and turn a crystal into a gas. Of course not all electron orbits are involved in producing the bonding waves. Only certain elements have electron orbits which possess the freedom to arrange themselves in the long rows necessary for magnetism.

Retarded Action at a Distance

Two radiating bodies can be traveling on straight parallel paths at the same speed and be occupying spots which are abreast of each other. Abreast of each other means that they occupy positions which are directly opposite each other. A line connecting the two bodies would be perpendicular to both paths and would represent the direct and shortest distance between the two straight and parallel paths. But, if the two bodies wished to signal each other, their two signals would not and could not follow the perpendicular line from one path to the other.

Retarded action means that messages or forces between particles always flow from the past position of the sender to the future position of the receiver. As long as forces move at finite speeds, all bodies will react to each other's moving presence with a delay proportional to the distances that their forces must travel. Moving presence means that bodies are aware of the changing position of another body and can adjust their position in a continuous manner but always with that delay caused by the time needed to become aware of a change.

Two bodies which are abreast of each other, as described above, and which are bonded to each other, as atoms are in a crystal lattice, will have two bonds, not one common bond. The path that the bonding wave travels in order to reach its target will be at an angle to the perpendicular direct line that would join them if they were resting. Two atoms moving on parallel paths will be joined together by two bonds that cross each other as the bonds flow along their angling paths.

A bonding wave which moves out at the speed of light, has no further connection with its atomic source.

A single body does not aim its force. The force is radiated. The sun's gravitational force, for instance, is moving in all directions at the same time. All the planets, all the meteors, all the space ships, the whole solar system, respond to this force. There are cases, however, when the force of two planets and the sun are aligned. The force of the three gravitational attractions can become stronger along that line.

The electrons in atoms radiate in all directions also. However, the manner in which electrons in orbits can maintain contact with nearby electrons in similar orbits, continually, over the entire 360° of every orbit, shows an organization, infinitely beyond that of the occasional alignment of two planets and the Sun. The movement of an electron around its orbit in an atom has been compared to the flow of an electric current around an electromagnet. Electron orbits create magnetic waves because the atom becomes a small magnet. Electrons radiate in all directions but an electron orbit is very directional. The north and south poles of this small magnet extend in the two directions which are perpendicular to the plane of the magnetic orbit.

In 1914, Leigh Page[16] suggested that radiation from electrons could travel as curved lines of force. Consecutively emitted elements of force which move away at the speed of light in a certain direction, may continue to travel in straight paths which are parallel to each other and moving in that same direction. The curves which are created by the electron's radiation in the two directions perpendicular to the plane of its orbit are spirals. One 360° turn of a spiral would occur in the same time as a 360° orbit of the electron. The distance along one turn of the spiral is the product of the speed of light times the time needed for the electron to complete one orbit. The elements of the spiral waves move in straight lines. They do not spin. The spiral waves move like a screw driven by a hammer, not like a screw driven by a screw driver.

Lateral Contraction

In the case above, where the two atoms are keeping abreast of each other as they follow parallel straight paths, the perpendicular line from each atom to the atom on the other path forms a right angle to each of the parallel paths. It is only when the crystal has no movement that the bonding force moves along this perpendicular path. The line of the bond that angles in the direction of movement will form the hypotenuse of a right triangle. The length of the hypotenuse, the length of the bond, is rigidly fixed by quantum conditions.

In the following discussion the length of the hypotenuse, the fixed length of the bond, the length along the diagonal are all one and the same and this length is given the value, 1. The time the bonding force needs to travel this distance can also be given the value, 1. The atoms, in their parallel paths, are moving at a fraction of the speed of light. This speed we can label V/C. As the time is 1 the distance traveled will also be V/C.

The length of the perpendicular that joins the two particles which are moving along the parallel paths is given the value, X. We thus have a right triangle with the length of the hypotenuse having the value, 1, the length of the side which is along the perpendicular to the two paths having the value, X and the side along the line of movement will have the value, V/C. The length of the perpendicular between the two paths, X, can be found by using the Pythagorean formula in which the hypotenuse squared less the square of the other known side will give the square of the length of the third side. The algebra is as follows.

$1^2 - V^2/C^2 = X^2$
$1 - V^2/C^2 = X^2$ Then give V/C the value 3/5.
$1 - 9/25 = X^2$
$16/25 = X^2$
$4/5 = X$

If we give the speed in the line of motion, V/C, the value, 3/5, when the length of the bond, the hypotenuse is 1, we will find that the distance between the parallel paths will become 4/5. The above algebra shows that, if the length of the bond is fixed, the dimension in the transverse direction will shrink as the speed, V/C, increases. The above line of thought started from the rather reasonable idea of retarded action at a distance. It is reasonable because the alternative is instantaneous action at a distance. Instantaneous action would mean that magnetic bonds would not be moving at the speed of light, and according to Maxwell's laws, laws which are respected by every physicist, magnetic waves move at the speed of light.

The above formula suggests that when bodies are bound together by electromagnetic bonds the actual or instantaneous distance between them will be affected by their relative motions. Contractions should also exist when the ties are gravitational because the bodies involved would be reacting to the moving presence of other bodies. Of course these bodies are not likely to be moving on parallel paths, while in a crystalline lattice, that could happen every time a line of motion coincides with the length of a row of atoms. The algebra above was an illustration of how and why a contraction could exist in the direction transverse to the line of motion. The motion does not have to be directly along the bonding line. A contraction can occur along a vector of the motion. The laboratory equipment in the Michelson-Morley experiment would definitely contract in the lateral direction if the lengths of the bonds are controlled by quantum conditions.

The theory that Lorentz settled on had one contraction in the line of movement which was the ratio of the contraction in the line of motion to the square root of that quantity, the contraction in the lateral direction. He had no contraction in the lateral direction. His choice was proven wrong by the results of the 1932 Kennedy-Thorndyke experiment. The Michelson-Morley experiment is being reviewed here for the reason that a two-contraction hypothesis is very much alive.

Chapter 14

The Michelson-Morley Experiment

When the Michelson-Morley experiment was first performed they were hoping that they could use the ether wind to measure the movement of the earth through the stationary ether. They expected to detect the difference between the speed of the earth through the ether and the speed of light waves through the ether. The ether was thought to be stationary and the breeze that we would detect would be similar to the breeze that we would feel when moving rapidly through still air.

A beam of light was split into two beams by a half-silvered mirror which allowed part of the light to be reflected at a 90° angle while the rest of the light passed straight through the mirror. After making the 90° turn, the L beam, the lateral beam, will travel a distance, d, to a mirror and will be reflected right back on that path to the same half-silvered mirror. A portion of the L beam will then go straight through that mirror to the interferometer.

The S beam, the straight-ahead beam, passed through the half-silvered mirror without being reflected, traveled straight ahead a distance, d, to hit a mirror and is then reflected right back on that path to that same half-silvered mirror. This S beam, on the return trip, strikes the back side of the half-silvered mirror, where part of the beam is reflected and joins the L beam as it enters the interferometer. Here the light is analyzed to determine how far the two rejoined waves were out of phase.

The whole setup was mounted on a platform that could be rotated. By turning the platform in every direction they were bound to find a place where a light path between mirrors would coincide with the direction of the movement of the laboratory through the ether. The round trip in the line of motion was expected to take more time than the round trip in the lateral direction.

Expected Time for Lateral Round trip

One method of illustrating the Michelson-Morley mathematics uses the following example. The time needed for a boat to cross the current of a river and return was contrasted with the time needed for a boat to move an equal distance up the current of a river and return. The boat, trying to reach a spot directly across the river, has to be aimed at a point that is upstream of that spot. The logistics of the problem would be about the same as if there were no current and the banks of the river were moving upstream.

In this problem they considered the distance directly across the river to be rigid. At the time the experiment was first run it had not yet occurred to them that the equipment of the laboratory could shrink due to its motion through space. Instead, the paths of the light waves were thought to be lengthened. The failure to detect the movement of the earth relative to the movement of the light waves was very puzzling and it was only after several years that a contraction theory was advanced.

The perpendicular distance across the river was considered to be one side of a right triangle and is called X here. The distance that the current flows while the boat is crossing the river is called Y and the diagonal distance that the boat travels, relative to the current, is called Z. The ratio of the distances, Y/Z, is equal to the ratio of the two speeds, V/C, because the current and the boat move through the same length of time. The algebra below involves the Pythagorean equation.

$$Y/Z = V/C, \quad Y = ZV/C, \quad Y^2 = Z^2V^2/C^2$$

The following is a standard Pythagorean equation and Z is the hypotenuse. The equality above is substituted for Y^2.

$$X^2 = Z^2 - Y^2$$
$$X^2 = Z^2 - Z^2V^2/C^2$$
$$X^2 = Z^2(1 - V^2/C^2)$$
$$X = Z\,(1 - V^2/C^2)^{1/2}$$
$$X\,/\,(1 - V^2/C^2)^{1/2} = Z.$$

This shows how much the diagonal distance, Z, increases over the direct distance, X, when the speed of the current, V, is a fraction of that of the speed of the boat, C. X, divided by a quantity which is less than 1, gives a value to Z, which is greater than the value of X. It was thought that the time for crossing the river would be increased by $1\,/\,(1 - V^2/C^2)^{1/2}$ when there was a current. The trip back would be affected by that same factor.

The mathematics of the Michelson-Morley experiment indicated that the time for the round trip in the line of motion would be enough longer than the time in the lateral direction that their equipment could detect the difference. We have just covered the situation that existed in the lateral direction. At this time we will go over the mathematics that indicated a time expansion for the round trip in the line of motion.

Using the analogy of travelling against the current and returning with the current, the speed would be slower going upstream, C – V. The speed would be faster coming downstream, C + V. The term, C – V, can be written as 1 – V/C and the term, C + V, can be written as, 1 + V/C. These speeds do not cancel each other because the time going at the higher rate of speed can be very much shorter. The time for the round trip in the line of motion can be calculated, using the following mathematics.

D is the distance traveled upstream and 2D equals a round trip. $D/(C - V) + D/(C + V)$ = time for round trip. In order to do this addition we must secure a common denominator. We can multiply the numerator and denominator of each fraction by the denominator of the other fraction.

$D(C - V) / (C + V)(C - V) + D(C + V) / (C - V)(C + V) =$
$D(C - V) / (C^2 - V^2) + D(C + V) / (C^2 - V^2) =$
$[D(C - V) + D(C + V)] / (C^2 - V^2)$
$[(DC - DV) + (DC + DV)] / (C^2 - V^2) = 2DC / (C^2 - V^2).$

This represents the total time needed for the light beam to make the round trip along the line of motion. To arrive at the distance traveled we must multiply by the speed of light, C. The round-trip distance is 2D multiplied by $C^2 / (C^2 - V^2)$. A number divided by a number which is smaller than itself is a fraction which is larger than one. This increase in distance should have been accompanied by a corresponding increase in the time needed for the round trip. They expected that the time for the round trip in the lateral direction would be lengthened by the factor, $1 / (1 - V^2/C^2)^{1/2}$, and the time for the round trip in the line of motion would be lengthened by the factor, $1 / (1 - V^2/C^2)$. It was expected that the two beams would be out of phase. The fact that the light waves were in phase with each other when they returned was, according to many writers, the origin of relativity theory.

The contractions which would have cancelled the excess time, as shown above, would have been the reciprocals of the above fractions. A fraction multiplied by its reciprocal would become 1. The reciprocal would be $(1 - V^2/C^2)^{1/2}$ for the lateral contraction. The reciprocal would be $(1 - V^2/C^2)$ for the contraction factor the in line of motion.

Chapter 15

Simple Arithmetic

There is a simple method of solving some problems in three variables. This is the method of putting the problem into a hypothetical form. We can illustrate the method by solving for the time needed for a light beam to travel a round-trip distance, moving with the motion of the earth in one direction, and against the motion of the earth in the other direction. This is the problem which you need to solve if you wish to understand the Michelson-Morley experiment.

The Michelson-Morley experiment involved a semi-transparent mirror, called a beam splitter, two regular mirrors, a source of light of one pure wavelength, and an analyzer which can discover if a light beam is still in phase. The analyzer is called an interferometer.

The source of light sends a beam of light to the half-silvered mirror. That mirror is set at a 90° angle to that beam of light. Part of the beam passes through the mirror and part of the beam is reflected so as to travel in the lateral direction. The reflected beam travels one unit of distance to a mirror which reflects the beam right back on the same path to hit the same half-silvered mirror. Part of this beam passes right through the half-silvered mirror and enters the analyzer which is in the other lateral direction.

A portion of the beam, which left the source to hit the half-silvered mirror, went straight through the mirror. This part of the beam also traveled one unit of distance to a mirror which reflected the beam right back on that same path to that same half-silvered mirror. Because the beam was now coming from the opposite direction, the angle of reflectance was in the opposite direction. The beam which made the round trip in the direct line of motion was reflected so as to join the beam which made the round trip in the lateral direction. They entered the analyzer together. The analysis showed that the two beams returned in phase with each other. There was no evidence that one beam had followed a longer path than the other.

The mathematics of the situation shows that the time for the light to travel the round trip in the lateral direction would be less than the time needed for the round trip in the direct line of motion. The scientific world marveled over what seemed to be a fact; light seemed to traverse different distances in the same length of time.

After some years, a rational explanation was advanced. It was suggested that the bonds holding the laboratory aparatus together, contracted just enough to offset the extra time that was needed for the light waves to travel the extra distances. This was not such an improbable theory because the bonds moved at the speed of light and we will find the speed of the laboratory through space will shorten the time of travel for these bonds also. The amount of the contraction would be the exact amount needed to offset the expected increase in time of travel for the light beams.

The mathematics of the contraction in the lateral direction was covered in chapter 14. It is hoped that it was presented so carefully that anyone with any curiosity about the universe could understand it. In this chapter the mathematics for the contraction in the line of motion is presented without the use of any algebra. For those who think that they can get by in arithmetic but refuse to even try to understand algebra, this is your chance. This book was written for you, also.

The following illustrates the hypothetical, no-algebra, approach to the problem. We can give the speed of light the value, 1, and give each lap of the round-trip distance, the value, 1. The test value for the speed of the earth in its orbit should be a very simple figure also, so that the arithmetic will be simple. A percentage answer can be applied to the real figures, or a formula can be verified. We will have no unknowns in the problem and no algebra will be needed.

The simplest value for the motion of the laboratory would be 1/2 the value of the speed of light. We can use 1/2 for the speed of the laboratory, 1 for the speed of light and 1 for the distance to the mirror and 1 for the return trip. The laboratory, moving with the earth, will travel 1/2 a unit of distance in 1 unit of time.

First, we wish to find out how long it will take a light beam to reach a mirror which is 1 unit away when the light beam leaves its source at the half-silvered mirror. The target mirror is moving away at half the speed of the light beam. For each unit of time the light beam will gain 1/2 a unit of distance. In two units of time the light beam will reach the mirror.

Now the light beam is 1 unit of distance from the half-silvered mirror. This is because the mirror and the source move with the laboratory equipment at the same speed. There is 1 unit of distance to cover, but the source is moving to meet the light beam at 1/2 the light beam's speed. The light beam will travel 2/3 of the distance and the mirror will travel 1/3 of the distance. For the light beam, units of distance are equal to units of time, and that time will be 2/3 of a unit on the return trip. The total round-trip time for the light beam would be 2 2/3 units.

You have been told that the light beams, in the Michelson-Morley experiment, returned as if there were no movement of the laboratory relative to the light beams. The waves came back as if there were no movement of the laboratory through space. They came back in 2 units of time, or so it seemed. The laboratory, however, had to be moving with the earth. This could be explained if the laboratory contracted just enough to conceal the extra time needed for the round trip. To calculate the contracted dimension, we can divide 2 by 2 2/3. Calculating would show that to be 75 percent. 6/8 = 3/4. The distance to the mirror and back would have been contracted to 75 percent of the path's length in order to explain the light beam's return in 2 units of time.

The light beam would be traveling twice as fast as the mirrors, so it would travel 1.5 units of distance in 1.5 units of time, while the mirrors are traveling .75 units of distance. On the return trip, the light would travel 2/3 of the distance and the source would travel 1/3 of the distance. Light travels twice as fast as the laboratory. 2/3 of .75 would be .5 units of distance and that could be traveled in .5 units of time. 1.5 plus .5 would give us 2 for the total time for the round trip. If the contraction occurred, the time for the round-trip distance would be 2. That shows that the contraction of the laboratory equipment could conceal the fact that the laboratory was moving relative to the movement of the light waves. Einstein abolished the relative speed of light rather than accept the possibility that the contraction was real.

In the preceding problem the contracted length of the laboratory apparatus was .75 of a unit. That can be written as (1 – .25) because .25 or 1/4 would be the amount of the contraction. When this problem has been worked out, using algebra, the answer for the contracted length is $(1 - V^2/C^2)$. When the speed of light, C, is given the value, 1, the speed of the laboratory can be written as V/C, a fraction of the speed of light. When the speed of the laboratory is given the value, 1/2, V/C would equal 1/2. 1/2 divided by 1 equals 1/2. V^2/C^2 would be 1/2 squared and that would be .25, or 25 percent. That would be the contraction in the line of motion. $1 - V^2/C^2$ or 1 – 1/4, would be the contracted length of the laboratory. Formulas can be very useful. We could square any speed of the laboratory and arrive at the amount of contraction that would then take place in the line of motion.

The contraction in the lateral direction is handled differently but the formula would still be simple. As was explained in Chapter 12, we would use the Pythagorean theorem for solving for the length of the hypotenuse of a right triangle. That would read, "The square of the hypotenuse is equal to the sum of the squares of the other two sides." We learned that in grade school.

Chapter 16

Some Reviewing

The very effective renormalization theory which was advanced by Schwinger, Feynman, and Tomonaga, and for which they received the Nobel prize, proved that Quantum Electrodynamics was not perfect in itself and was open for some basic improvement. Three books are quoted on this subject. The first book is "QED," (Quantum Electrodynamics), by Richard Feynman[48]. He calls "renormalization," a dippy process and refers to it as resorting to hocus-pocus. He was referring to the manner in which they removed infinite as the weight of an electron and put in the real weight.

The second book is by Eric J. Lerner[46], "The Big Bang Never Happened." He explained that quantum electrodynamics treats electrons as infinitely small mathematical points Since electrical force increases as distance decreases, as in the inverse square law, the electron has an infinitely high energy to go with its infinitely small size. Like charges repel each other. The electron is all one like charge, therefore the electron should explode. QED predicted that the electron had infinite mass. Lerner says that was nonsense. An electron has a mass of 10^{-27} gram. It wasn't that the physicists didn't see this contradiction. They just ignored it. When the real mass of the electron is placed in their equations instead of that theoretical mass, they had a theory that worked. That corrective procedure was called, renormalization.

The third book, "Genius, The Life and Times of Richard Feynman[49]", by James Gleick, had much to say about renormalization theory. One passage, which illustrates the reservations which Feynman and other physicists held, is worth quoting here. "From one perspective, renormalization amounted to subtracting infinities from infinities, with a silent prayer. Ordinarily such an operation would be meaningless. The infinity of all the odd numbers, subtracted from the infinity of all the real numbers would still leave the infinity of all the even numbers." "The theorists hoped, when they wrote that infinity minus infinity would still equal zero, that a miracle of nature would make it so, for once. That their hope was granted said something important about the world. For a while it was not clear just what." Quantum Theory was then able to justify the rules for chemical structures and chemical bonding. In this author's opinion they just gave Quantum Theory a good dose of common sense.

Ether

When two orbital planes are adjacent to each other, the south wind of one will blow through the north end of the other, and the north wind of the other will blow through the south end of the first. The orbit of an electron will turn at the same speed as the turn of the helical wave. If three orbital planes are spaced one wavelength apart, the first helical wavelength can make a three-way intersection with the orbit of the middle electron and a wave moving from the opposite direction, from the third electron.

The electron travels around the atom in its orbit in the same amount of time that the wind needs to carry the wave from one orbital plane to another. The electrons move at a slower speed than the waves, therefore the electron's path must be much shorter. As the electron moves around its orbit the wind picks up a unit of pull as each diameter of the electron crosses the path of the wind. The pull of the electron on another electron has been likened to a shadow or the pull of a vacuum. If an electron and a wave cross each other at a right angle, the wave will exercise one unit of pull on the electron. If a wave follows an electron completely around its orbit the wave can exercise ten thousand diameters or units of pull on the electron. At the same time the wave will be picking up additional units of pull that can be exercised if additional orbital planes are encountered.

When two electron orbits are in planes, parallel to each other and one wavelength apart, the arrival of each wave can be in phase with the arrival of the electron in the adjacent orbit. This means that the pull of each wave follows the other electron at the shortest distance possible, completely around its orbit.

4 pi r^2 is the formula for the expanding surface of spherical radiation. The radiant forces which control the orbits are divided along the length of a 2 pi circumference and weaken inversely with the square of the distance to a point on that circumference, In atomic research h is divided by 2 pi. That can be a division of the energy in one orbit. That could give us the angular velocity for every point on the length of that orbit. The inverse square law developed very naturally from the inverse r^2 above. Electromagnetic waves are directional and result from magnetic orbits and electrical spins. Gravitational waves are not directional, they are radiant.

This ether has many useful features. The square of the speed of light times the mass of all the ether particles allows the development of a measurable momentum. The radiant nature of the ether explains all kinds of radiation. The very small size of the ether particles allows them to discover the smallest interstices of electrons or other matter and there deliver momentum. The small size of the ether particles explains the ability of remnants of light waves to travel billions of light years to end up disturbing an electron on the surface of a photographic plate.

There were many ether theories in the nineteenth century. For the most part they theorized about stationary ethers or jelly-like ethers. The Michelson-Morley experiment was supposed to reveal the earth's motion through a stationary ether. When that experiment failed to show that movement, the general reaction was that all ethers were ruled out. That was a measure of how little favor the energetic LeSage ether had at that time.

The Fitzgerald-Lorentz contraction hypothesis, which seemed to save the ether, was criticized for being too flimsy, too "ad hoc". Faith in ether theories lost favor. When Albert Einstein suggested that we forget all about ether theories, his suggestion was taken as a command by most of his followers.

Generations of physicists, and the journals they publish in, have avoided that subject. There is great loyalty to Einstein's relativity theory. It seems that only a historian will mention the usefulness of ether theories. Einstein did seem to relent later and said that an ether theory might be necessary, but by that time the quantum theory had taken over with its ban on the possibility of the existence any kind of reality or causes or structures below our present knowledge.

The biggest mysteries of the universe are left unexplained. The physicists ignore these problems as if they were already solved. There seem to be spoken and unspoken hypotheses that some things just happen to be in the nature of things and therefore need no explanation. Of all the mysteries that an ether theory should deal with, the biggest one would be the how and where of the immense store of energy that seems so readily available for all the phenomena involving motion and heat.

In earlier times, physicists found many clues on which to base ether theories. In free space, light waves move in straight lines. The gravitational force between two bodies is measured along the straight line that would connect their centers. Momentum moves mass in straight lines until that mass meets a force that can change that course. The straight line should have a place in any theory of an ether.

George LeSage, a Swiss physicist hypothesized that very small ether particles, moved in straight lines, in every direction, at an enormous speed. The particles of matter would be pressured from every direction by the much smaller ether particles. If two particles of matter partially shielded each other from the constant bombardment, each would have less pressure on the side nearest the other. They would be pushed toward each other.

In 1873, in an article in the Encyclopaedia Britannica, Clerk Maxwell[28] criticized LeSage's energetic multidirectional ether theory by saying that the action of such an ether would raise the temperature of the universe to a white heat and raise all bodies to an enormous temperature. He was saying that this heat would be due to the pressure caused by the enormous bombardment.

Maxwell thought that this ether would cause heat, but there are other ways to look at the problem. LeSage's ether would produce motion and only the motion would produce heat. The vibrations of heat within molecules and atoms could arise only if the parts moved relative to each other. There could be no vibrations and no heat waves without this relative motion. The ultimately smallest particles of matter would be subjected to the pressure of the LeSage ether. If these smallest particles have no parts to vibrate and no apparatus to create heat waves, how could we be aware of any heat caused by the pressure on them? It is a generally accepted or common knowledge among scientists that heat is motion. The motion is that of electrons and other particles of matter. There was once a theory that claimed that heat was a substance called caloric. This was dropped before Maxwell's time.

The ether particles are a very unknown quantity. How was Maxwell so sure that the ether particles would cause heat before they cause motion. The multidirectional ether, by definition, moves in every direction at the speed of light. The ether that could strike a smallest particle of mass within one second's time could come from points as far away as the surface of a sphere, which has a radius of 186,000 miles. The ether particles which strike from a particular direction may follow each other at intervals of hundreds of miles or in a fraction of a second.

Maxwell's, 1873 criticism was very widely read because it was reprinted in other editions, up to and including the 1890 edition of the Encyclopaedia Britannica. Maxwell, however, did not kill the LeSage ether theory. Aronson[29] tells of S. Tolver Preston and Sir William Thomson working hard to revive the LeSage theory in the 1880s. S. Tolver Preston[11] published two editions of a full length book describing the many advantages of LeSage's theory in explaining gravitation. Small changes could meet lots of arguments. For instance, the ether particles may be magnetic.

Whittaker, in his history, told of Sir J. J. Thomson, in 1907, Ebenezer Cunningham, in 1914 and 1915, and Leigh Page, in 1914, using the LeSage ether to explain the lines of force in energy fields[18]. They were theorizing that electrons create waves in the ether winds which would form fields of force.

The main evidence for the radiant nature of the ether lies in our radiant universe. This radiant nature is expressed in many ways. Gravitation furnishes one illustration. The gravitational attraction of every single particle seems to reach and influence every other particle in its vicinity, no matter in what direction. The manner in which the earth's gravitational field seems to exist at every spot in the path of the moon, is part of the evidence. The gravitational radiation emitted at any single instant must be spherical or it would not have the ability to extend in every possible direction at the speed of light.

Light is another example of a natural force moving in every possible direction. Light from billions of stars can converge on the small eyepiece of a telescope. Every speck of free space seems to be a radiant center where light flows in from all directions and out in all directions. Light from a galaxy a billion light years away will pass through billions of such radiant centers every second of its journey.

One might ask how such a thing could be and a possible answer is as follows. The light photons are composed of groups of very small particles, or they are composed of groups of holes in a moving, radiant ether. They have to be very small to have such a long free path. This refers to the average distance that a particle could travel without hitting another of the moving particles. If the size of the particle is 1/X the length of the mean free path will be X^3.

The mathematics would be somewhat as follows. A one inch diameter ball in one inch square box has the freedom to stay put. In a two inch square box it has eight times as much room. In a three inch square box it has twenty-seven times as much room in which to travel. The effect would be similar if the balls became smaller and the box retained its size. The LeSage ether theory, as proposed by S. T. Preston[31] would obey the gas laws with the entire universe as the container.

The electron's field is also radiant. Its magnetic field extends in every direction. When electrons flow down a wire a moving magnetic field completely surrounds the wire. The well-known iron filings test shows that a magnetic field completely surrounds the magnet. We know that the electron radiates to the gravitational field because the electron has weight. That the electron radiates in all directions is accepted by scientists[17]. Endlessly and forever every particle of mass continues to radiate that field force called gravitation. Endlessly and forever each proton continues to radiate an electrical attraction which is also a pull. An electron and a proton can pull on each other and the size of that pull varies with their distance apart.

Einstein, over the last decades of his life, had a consuming ambition devoted to discovering the relationship between gravitation and electromagnetism. Einstein and other physicists thought that would he a matter of supreme importance if it were only possible. To this day the physicists have not discovered a, "Unified Field Theory."

Einstein, at the time he invented his fourth dimension and abolished the relativity of the speed of light, delivered a third blow to common sense by effectively abolishing the ether. He had no need for an ether theory, and he said so.

In the nineteenth century physicists thought that the first step toward understanding the field forces would consist of forming a theory of an ether. Magnetic and electric forces gave great evidence of being waves in an ether. The nineteenth century theorists thought that they needed an all-pervasive ether in which those waves could wave.

This author, by disregarding the prohibitions of relativity theory and by following the laws of algebra and the laws of common sense, and by following a unified field theory which he would prefer to call a paradigm, has discovered independent solutions or explanations for a tremendous number of experimental results, which could also be called facts or numbers.

About thirty-five years ago this author decided that there was overwhelming evidence for the existence of an ether for the transportation of light waves and every other kind of wave. There has to be something for these waves to wave in. This medium must consist of winds of small particles which blow through each other in all possible directions at the speed of light. A wind of small particles which passes perpendicularly through the plane of an electron orbit, will suffer a helical path to be cut through its directional density.

This author had worked with this ether about a year when he discovered that a very similar theory dated back to 1744 and originated with a Swiss physicist, named George LeSage. LeSage was still fighting to get his theory accepted when he died in 1803. In 1883 S. Tolver Preston had developed LeSages theory into a full-length book.

In 1996, this author is not a Don Quixote, tilting with windmills. This is for real. This book describes a common-sense universe which will replace the uncommon-sense universe which relativity theory and quantum theory describe. Many millions of persons can understand the simple algebra that is used in describing this common-sense universe, when this common sense is made available to the millions.

As far as the physical sciences are concerned and as far as the defense of logic is concerned this may be the most important book published in this century. The assault on common sense and logic by the frankly unashamed attacks by relativity and quantum theories has eroded support for all kinds of critical thinking.

Lorentz did not realize that the Michelson-Morley experiment could be explained by a two-contraction theory. The mystery of the hydrogen orbits and the planetary orbits can be explained now by a solution for that greatest of all mysteries, momentum at less than the speed of light. It is such a simple theory too. Magnetism can be explained by double-helical magnetic bonds. There is a now a gravitational theory that just might explain Mercury's perihelion. There is a new Bode's law for the planetary orbits. The first real change in several hundred years.

The physicists developed theories which are based on proven physical facts. It is no wonder that the facts fit the mathematics of their theories because the mathematics was chosen to fit their interpretations of those facts and/or the interpretations were chosen to fit the mathematics they had invented.

Glaring inconsistencies in their theories are freely admitted by the physicists. They can explain them all by the statement that this is not a common-sense universe. If you insist on a common-sense universe you are labeled as being naive and ignorant of the fine points of modern science. They will point out how well the mathematics fits their theories.

Chapter 17

Contraction Evidence

The explanation for the contraction in the line of motion is complex and involves the inverse-square law as it affects magnetic bonds. The results of the Michelson-Morley experiment indicate that mass contracts in the line of motion and that the contraction is proportional to the formula, $(1 - V^2/C^2)$. Compelling reasons have been found for believing that this amount of contraction is real. The inverse square law says, that as a bond is extended, the bond becomes weaker. The converse, is of course, also true. When the same bond is shortened it becomes stronger.

The interlocking mathematical evidence goes far beyond any possibility that its existence is due to contrived or accidental coincidence. The mathematics of the contraction itself is covered in the drawings and in the explanations of the drawings. The contraction is due to the necessity for maintenance of phase relationships along the bonds that hold matter together. Most of these clues for the bonding waves in matter have come from the study of magnetic waves. It is in magnets that the bonding waves can become so strong that they can emerge into the open air and thereby reveal their secret structure.

To continue this discussion, a historical reference is repeated. Sir J. J. Thomson[3], who is credited with discovering the electron in 1898, did a lot of theorizing about magnetic waves. He thought that there was a fan-like mechanism within a magnet which created rotating waves and sent them in opposite directions at the speed of light. Thomson did not say that the waves were spiral but it follows from the situation that they would have to be. A wave which spins around its direction of motion must be spiral. An electron which spins around the direction of its atom's motion must also be following a spiral path.

Basic physics will teach you that electrons, in magnetic orbits, are the source of magnetism. Long rows of magnetic atoms join together in producing strong waves that emerge from the north and south poles. It is a cardinal principle of wave action that waves in phase can reinforce each other and waves out of phase can destroy each other. If the hills on one wave join with the valleys on another wave they will cancel each other.

An electron, in a magnetic orbit, is the fan-like mechanism that Sir J. J. Thomson was suggesting. The two waves that flow in opposite directions are the waves that emerge from the two poles of the individual atomic magnets. The magnetic forces which flow through long rows of atoms in order to produce the strong waves, must be flowing through each other in both directions. We should have no trouble in believing that the magnetic waves can pass through each other, After all, light waves pass through each other to reach us from the most distant parts of the universe. They pass through other waves, millions of times a second, for trips which last billions of light years.

The magnetic waves move at the speed of light in the two directions which are perpendicular to the planes of the electron orbits. The spiral waves will travel without rotation, like a screw driven with a hammer, not like a screw driven by a screw driver.

Two spiral waves, sharing the same tube of space, but traveling in opposite directions, will be in continuous intersection with each other and with electrons in magnetic orbits. The opposite waves will not affect each other except through their affect on the electron.

The electrons in their orbits are held locked in place at the intersections of the oppositely moving helical waves. If the pulls on the electron are uneven, the stronger pull will create the movement that will restore the balance. The imbalance in the pulls may be cured in several ways. A continuous imbalance may cause continuous movement. Balance may be also restored by contractions or expansions or by sudden releases of energy.

In the forward direction the wave will have to catch up with an orbitting electron which is moving away. In the reverse direction the wave will be meeting an orbitting electron. The inverse square law, which applies to magnetic waves, as well as to gravitational waves, will say that the pull of those waves will vary inversely with the square of the distance they have traveled from their source.

We have shown that retarded action at a distance, as applied to the atomic bonds in a crystal, would cause a crystal to shrink in the direction lateral to its line of movement. The laboratory equipment for the Michelson-Morley experiment was held together by atomic bonds and therefore a contraction should have taken place due to retarded action at a distance. Michelson and Morley were thinking that the transverse distance across the equipment would be rigid and that the path of the light wave would be lengthened and forced to travel in the diagonal direction. The length of the bonds should be controlled by quantum conditions. The length of the bonds should change in jumps, not gradually.

It was shown in chapter 12 that the perpendicular distance between two parallel paths would shrink. The amount of shrinkage would be to the square root quantity, $(1 - V^2/C^2)^{1/2}$. This is the reciprocal of the amount of the expansion of time used, which was anticipated by Michelson and Morley. As the contraction in the lateral direction this would help explain the results of the Michelson-Morley and the Kennedy-Thorndyke experiments.

The explanation of the contraction in the line of motion is an entirely different matter. This explanation involves the contraction of mass in motion, the rotation of the double helix, the inverse-square law, the creation of momentum, the explanation of the many aspects of magnetism, the explanation for the results of the Michelson-Morley experiment and the explanation for the results of the Kennedy-Thorndyke experiment. The most important item, however, is the manner in which the three drawings can show how phase is maintained along long rows of magnetic orbits. The reason for using three drawings instead of one is the desire to show that phase is maintained and the contraction formula is maintained despite the changes in the speed.

In the sixty-some years since the Kennedy-Thorndyke experiment was performed, the history of science has been very silent on the possibility of two contractions as a solution to the mysterious results of the Michelson-Morley experiment. A 1937 paper by H. E. Ives[42] is one exception. He showed that the results of both of the above experiments would have had to have been null if the two contractions occurred. If there were a contraction in the line of motion of $(1 - V^2/C^2)$ and a contraction in the lateral direction of $(1 - V^2/C^2)^{1/2}$, the movement of the earth would never affect the time for the travel of the light beams in the Michelson-Morley experiment. The results would always be null.

To this author it seems unfortunate that Lorentz chose one contraction instead of two contractions in his effort to explain the null results of the Michelson-Morley experiment. The transformation equations would not have been written as they were, and that could have changed the course of history.

An electron radiates elements of force in all directions. As long as it is unorganized radiation, no direction is favored. When electrons are in orbits, however, the forces can be very directional. The organized forces travel in two opposite directions in lines which are parallel to each other and are perpendicular to the planes of the magnetic electron orbits. These directions would be directly left and right in the drawings. It may be argued that the electron orbit has so much movement in the forward direction that it doesn't orbit in a plane. Actually almost all the movement to the right in the drawing is due to the movement of the atom to which the electron is attached.

In answer to that I will first tell you how many orbital wave lengths there are in one second. That would be 4.566913×10^{14} orbits per second. That would be 2.997174×10^{14}, meters per second, the speed of light, divided by 656.28×10^{-9} meters which is the wave length of the red spectrum of the hydrogen atom. If the movement were at 1/1000 the speed of light, the orbit would much more resemble movement in a plane.

Electrons are not labeled but will always be at the tip of the spreading wavefronts at the latest time shown. Two waves spread out behind an electron like waves spreading in two opposite directions behind a racing motorboat. When elements of force are emitted at successive instances from an electron in its orbit there will be a regular increase in the distance that an emission has traveled from an electron's path as we go back in time along that path. If, on the drawing, you measure both left and right, from where the electron was in its orbit, you will find the distances that the waves have traveled will be the same.

When electrons are orbiting, the elements of force which are sent out in parallel and perpendicular directions, at successive instants, will form curved lines of force, moving spirals, which function as the walls of a tube-like space. The diameter of the tubes would be the diameter of the electron orbits. These electrons will orbit in the walls of that same tube of space. The best evidence we have for the existence of orbits is the existence of wave lengths.

The curved line of force will be very thin. There is the factor of the tiny size of the electron. It is reasonable to assume that the line which it would produce would be very thin. The curve of the line of force must have the same shape, orbit after orbit. One section of a wave will follow millions of electrons around their orbits as it progresses from one end of a magnet to the other. This is how magnetic waves are able to accumulate the force which emerges from the poles of the magnet.

The force of a magnetic wave is a pull. The repulsion which magnets seem to have for each other in certain orientations does not prove otherwise. Two magnets can simply pull themselves away from each other when two north poles or two south poles are set to face each other. The fields that project from each magnet into the other magnet will be mingling with the field in the other magnet but the two fields will be turning in opposite directions. The three-way intersection depends on the opposite waves turning around the orbit at the same speed. When the wave slows around the orbit because of the interference, the pull it could exert would not be able to meet and cancel pull from the other direction. Momentum would be created in each magnet to pull them apart.

Two magnets can pull themselves together when north poles are set to face south poles. That this *is* the natural orientation of the poles within the magnet can be easily proved. Cutting magnets in two will show that north poles always face south poles. This supports the idea that magnetic waves act as bonding forces within magnets. The force which pulls them together would not disappear as soon as they are together.

Chapter 18

Explanation of Drawings

The drawings illustrate the contraction of mass due to motion as it occurs along a double helical wave system within a magnet. The helical paths of the waves and of the electron orbits are shown as slanting lines on flat paper. Simply show as helical by rolling into a tube 360° in circumference. Degrees are shown at the right. Atoms are not shown, just the paths of the magnetic orbits which are attached to atoms.

The dashed lines which move to the right are the helical wave fronts which must overtake intersections with electrons which are also moving to the right. The dotted lines which move to the left are the helical wave fronts which are meeting intersections which are moving to the right. The dotted lines and the dashed lines are moving lines, however, they show no movement in the drawing. It is as if their motion were frozen by the speed of a very fast camera. There is only the one time along all their lengths. That time is the time shown at the three way intersections at the apex of the V of the spreading waves, the point where the waves are just beginning to spread from the three-way intersections.

The letter, C, represents the distance which a wave can travel while an electron is making a 360° orbit. The letter, C, also reflects the fact that the waves are moving at the speed of light. C is the length of the waves when there is no movement of the atoms which control the orbits which create the waves and when there is no motion along a line which would give the electrons, in their orbits, spiral paths. The speed of the magnet or the system, is labeled, V. This speed is always a fraction of the speed of light. When C is given the value, 1, the speed of the system can be written as V/C. Then C - V becomes 1 - V/C and C + V becomes 1 + V/C. When V/C - 1/4, the contraction to $1 - V^2/C^2$ translates to 1 - 1/16. $(V/C)^2$ is the amount of contraction in the line of motion.

When V^2/C^2 equals 1/16, we will need to use 16 spaces in order to diagram how one space overlaps every 16th space. This is leading up to the reason why we are using such improbably high speeds. If V/C were equal to 1/1000 C, $(V/C)^2$ would be 1/1000000. We would need a million spaces to illustrate the same situation. Even V/C = 1/8 would crowd the page beyond the tolerance point. Three different drawings are offered in order to show that the contractions occur, regardless of the speed of the bonded systems.

When V/C equals 1/4, the spaces in the vertical direction are 1/4 times 360° or 90° each. When V/C equals 1/3, the spaces in the vertical direction will be 1/3 times 360° or 120°. V^2/C^2 would equal *1/9* and we will need 9 spaces in the horizontal direction.

There is a disadvantage in using, 1/2, 1/3, and 1/4 as hypothetical values when discussing the l - V/C relationships. Those values cannot, at the same time, be used to illustrate the creation of momentum. This is due to the distortion of the momentum equations when speeds approach the speed of light. The value, .0001 C which was used in Chapter 2, would show the exactness of the creation of momentum to the eighth decimal point. That speed was about 300 kilometers per second. Using the value .000001 C, which would be about 3 kilometers per second, the exactness would be to the twelfth decimal place. The length of a wave will be the distance it travels from its point of origin on an electron to the point where it meets a wavefront coming from the opposite direction. This is the point of a three-way intersection. We cannot say that the length of a wave is the distance from one three-way intersection to the next one. That would not be true. In the case of the shorter waves to the rear, you can notice that they end at a three-way intersection but do not start at one. Check the drawing for V = 1/4 C, by counting the spaces, you will notice that the wave which leaves electron A, at time, t1, will reach a three-way intersection with electron B, at time, t6, after traveling 20 spaces. The wave in the forward direction is 5/4 C, and the contraction is 1/16 C. 20/16 C. will equal 5/4 C.

At time, t3, a wave leaves electron C and travels 12 spaces to the left to the three-way intersection at Bt6. 12/16 C will equal 3/4 C. 5/4 C and 3/4 C will equal 2 C. Note that two waves will always add up to 2 C on the straight line where the pulls on the electrons at the intersections contain the factors, C – V and C + V. The contractions do not affect that relationship. Electron C had already traveled 180° from its last intersection at time, t3. Electron C will travel another 270° before the wave which left electron C at time t3 will reach the intersection with electron B at time t6.

At time, t6 there will be a two-way intersection on the far side of the tube. Note where the wave-fronts are crossing between the orbits. At a two-way intersection the helical waves have not the slightest effect on each other.

If you will inspect the diagrams of the contractions, you will see that, in the forward direction, the first wave starts at At1, the second wave starts at Bt1, and the third wave starts at Ct1. Each wave starts one space before the wave above it meets its three-way intersection. This overlapping is the exact amount of the contraction.

Along a long row of orbits each orbit is between two immediately adjacent orbits. We can call three orbits, A, B, and C, and say that the magnet is moving to the right. You can pick any three orbits at random along thousands or millions of magnetic orbits and the same situation will apply.

The orbital paths of the electrons slant up the page and to the right as time flows up the page and to the right for the electron. Along the wavefront shown as a dotted line, time flows, to the left. A1ong the wavefront shown as a dashed line, time flows to the right.

The three-way intersections are the points where electrons are held in the grip of waves from two directions, at the same time that they are adding increments of force to these waves. The waves which move away at the speed of light can have no further influence on electrons which they left, behind.

Along a long series at atoms with electrons in magnetic orbits every electron will be at the intersection of waves from opposite directions. This assures that wavefronts will always be in phase. A wave will become stronger and stronger as it picks up strength at every orbit that it passes through. It is in this way that magnetic waves become strong enough to emerge from the ends of a magnet. At the same time that electrons are adding strength to two waves, they are being held tightly in their orbits by those same two waves.

There is a point where the loss of strength, that goes with distance traveled, cancels further increments of force and the interior strength of the magnet levels out. Each increment will lose strength according to the inverse-square law. It also explains the fact that the interior strength of a bar magnet is at one half strength just inside the two poles and becomes double strength over most of the length of the magnet. The magnetic wave which emerges at each pole is at full strength just before and just after it emerges. The wave which begins at that pole and emerges at the other end will start at almost zero strength. V is the speed of the earth at the point where it is carrying the Michelson-Morley laboratory through space.

V/C is the speed of the laboratory expressed as a fraction of the speed of light.

V/C can be used to express proportional distances as well as proportional speeds. The following formulas could also be considered to be percentages as compared to measurements when V = 0.

V^2/C^2 is the amount of contraction which occurs along the line of motion in the Michelson-Morley experiment.

$1 - V^2/C^2$ is the total contracted length of the light path when it lies along the line of motion of the laboratory. $(1 - V^2 / C^2)^{1/2}$ is the total contracted length of the light beam when its path lies along a line lateral to the line of the motion of the laboratory.

The following apply to two-way magnetic bonds between adjacent atoms in a substance moving at speed V with bonding waves moving at the speed of light, C.

C – V is the distance which a magnetic wave travels when, in the line of motion, it is pulling on an electron orbit which follows it.

C + V is the distance which a magnetic wave travels when, along the line of motion, it is pulling on an electron orbit which it is following.

$1 / (C - V)^2$ is the value of the inverse square pull which a magnetic wave exercises on an electron orbit which follow it. This is a pull in the forward direction.

$1 / (C + V)^2$ is the value of the inverse square pull which a magnetic wave exercises on an electron orbit which it follows. This is a pull in the rearward direction.

$1 / (C + V)^2$ subtracted from $1 / (C - V)^2$ always gives a surplus of pull in the forward direction. That surplus determines the amount of continuing momentum which resides in those bonding waves. At normal everyday speeds the amount of pull is 4 times the numerical value of the speed, V, As speeds approach the speed of light much more pull is needed to increase speeds.

References

1. J. D. Watson, "The Double Helix," Signet Books, New York, 1969, p. 38.
2. J Clerk Maxwell, "Encyclopaedia Brittanica," Tenth Edition, article, "Atom," p. 46.
3. Sir J. J. Thomson, "Recent Researches in Elect. and Mag.," Clarendon Press, 1893, p. 35.
4. Max Planck, "Where is Science Going?," Ox Bow Press, Woodbridge, Cn. 1981, pp. 44-5, 178.
5. Thomas Kuhn, "The Structure of Scientific Revolutions," New American Library, New York, 1986, pp. 30-1, 78.
6. Martin Gardner, "Relativity for the Millions," Pocket Books, New York, 1965, p. 69.
7. Alfred North Whitehead, "Science and the Modern World," New American Library, New York, 1925, p. 48.
8. Sir Edmund Whittaker, "History of the Theories of Aether and Electricity," Harper, New York, 1960, Vol. 1, p. 317.
9. Cantor and Hodge, Editors, "Conceptions of Ether," Cambridge Univ. Press, New York, 1981, p. 18.
10. "Conceptions of Ether," Larry Laudan, pp. 164-9.
11. S. T. Preston, "Physics of the Ether," Spon, London, 1875, pp. 1-23, 115.
12. "Whittaker's History," Vol. I, p. 190.
13. "Conceptions of Ether," Daniel Siegel, p. 245.
14. "Conceptions of Ether," Daniel Siegel, pp. 246-7.
15. "Whittaker's History," Vol. I, p. 301.
16. Leigh Page, "Relativity and the Ether," Am. J. Sci., Vol. 38, 1914, pp. l69-187.
17. "Whittaker's History," Vol. II, p. 36.
18. "Whittaker's History," Vol. II, p. 247-9.
19. Claudio Rebbi, "Solitons," Scientific American, Feb. l979.
20. Ruggero H. Santilli, "Il Grande Grido," Alpha Publishing, Newtonville, Ma., 1984.
21. Paul Gerber, "Die Fortptlanzungsgeschwindigkeit der Gravitation," Ann d. Physik, Band 52, 1917, pp. 415-451.
22. Demys, Scientific Ethics," Cambridge, Mass. 96 Prescott St. Vol. l. No. 5, 1985.
23. Werner Heisenberg, "Encounters with Einstein," Princeton Univ. Press, 1989, pp. ll3-122.
24. Petr Beckman. "Einstein Plus Two," The Golem Press, Boulder, Colo. 1987, pp. 62, 166.
25. Petr Beckman, "Einstein Plus Two," p. 26.

27. Rudolph T. Lee, "The Universal Mechanism," Lee Publishing, 1959, Baywood Park, Calif. pp. 80-81
28. J. Clerk Maxwell, "Encyclopaedia Brittanica," Ninth Edition, Article "Atom," pp. 41-2, or pp. 46-7.
29. S. Aronson, "The Gravitational Theory of Georges Louis LeSage," in Natural Philosopher, 1964, p. 59.
30. J. Clerk Maxwell, "Encyclopaedia Brittanica," Ninth Edition, p. 41 or p. 46.
31. S. T. Preston, "Physics of the Ether," Spon, London, 1875, p. 14-32.
32. "Whittaker's History," Vol. I, p. 196.
33. "Whittaker's History," Vol. I, pp. 191-1.
34. D. S. Parasnis, "Magnetism, from Lodestone to Polar Wandering," Harper, New York, 1961 pp. 25-28.
35. "American Institute of Physics Handbook," McGraw-Hill, New York, 1957, #5 p. 240.
36. L. Silberstein, "The Theory of Relativity," Macmillan, London, 1914, p. 79.
37. R. W. Clark, "Einstein, The Life and Times," Avon Books, New York, N. Y. p. 127-31.
38. R. W. Clark, "Einstein, The Life and Times," p. 526.
39. J. M. Bird, "Einstein's Relativity and Gravitation," Scientific American Publishing Co., New York, 1922.
40. "Encyclopaedia Brittanica," 1960, Vol. 12, p. 471, under interferometer, Vol. 15, p. 417, or Michelson-Morley Exp.
41. R. J. Kennedy and E. M. Thorndyke, "Experimental Establishment of Relativity of Time," Phys. Rev. 42, 401, 1932.
42. H. E. Ives, "Graphical Exposition of the Michelson-Morley Experiment," J. Opt. Soc. Am. pp. 27, 177, 1937.
43. Guy Murchie, "Music of the Spheres," Riverside Press, 1931, Cambridge, Ma. pp. 460-80.
44. Gerhard Herzherg, "Atomic Spectra and Atomic Structure," Dover Publications, 1944, New York, N. Y. p. 25.
45. Sir James Jeans, "The New Background of Science," Macmillan Co., 1934, Cambridge, England, pp. 55-62.
46. Eric J. Lerner, "The Big Bang Never Happened," Vintage Books, New York, N. Y. pp. 33-4, 53-4 350.
47. Ruggero M. Santilli, "Il Grande Grido," Alpha Publishing Newtonville, Mass. pp. 1-8,
48. Richard P. Feynman, "Q E D," Princton University Press, Princeton, N. J. p. 128.
49. James Gleik, "Genius, The Life And Times of Richard Feynman," Pantheon Books, New York, 1992, 239-240.
50. David Bohm, "Causality and Chance in Modern Physics," Harper Torchbook Edition, New York, N.Y. pp. 85, 126

51. Isaac Asimov, "Understanding Physics III," George Allen And Unwin, London, 1966, pp. 73-77.

THE CONTRACTION OF THE MAGNETIC DOUBLE HELIX
Drawing 1 When V = ½ C

The value, C is the distance a wave will travel and the value, V, is the distance that the magnet will move, while the electron is completing one orbit. A wave's length is measured from its origin on an electron to its meeting with a wave front from the other direction. The helical paths of the wave fronts and electrons are shown on flat paper. Slanted paths on paper become helical when the paper is rolled into a 360^0 tube. Degrees are shown on the right. The nuclei are not shown. The paths shown are only those made in the period, from time, t1, to time, t4. The paths of the electron orbits, from time, t3, to time, t4, are shown twice. These are identical positions 360^0 around the tube. $1 - V^2/C^2$ is the formula for the contraction in length in the line of motion. When V/C is ½, V^2/C^2 will be ¼. That will be the amount of the overlap and the size of the contraction. The length of a forward moving wave is 3/2 C or 6 spaces. The length of the reverse wave is ½ C or 2 spaces. C = 4 spaces. At time, t4, electron B will arrive at the junction of waves that left electron A at time, t1, and left electron C at time, t3. The electrons are held firmly between the opposing pulls of the two waves as the intersection circles the tube. The right flowing waves, the forward waves, are shown as dashed lines. The reverse waves are shown as dotted lines. The spiral paths of the electrons are shown as solid lines. There are more details in the text.

The Contraction of the Magnetic Helix
Drawing 1 When V = ½ C

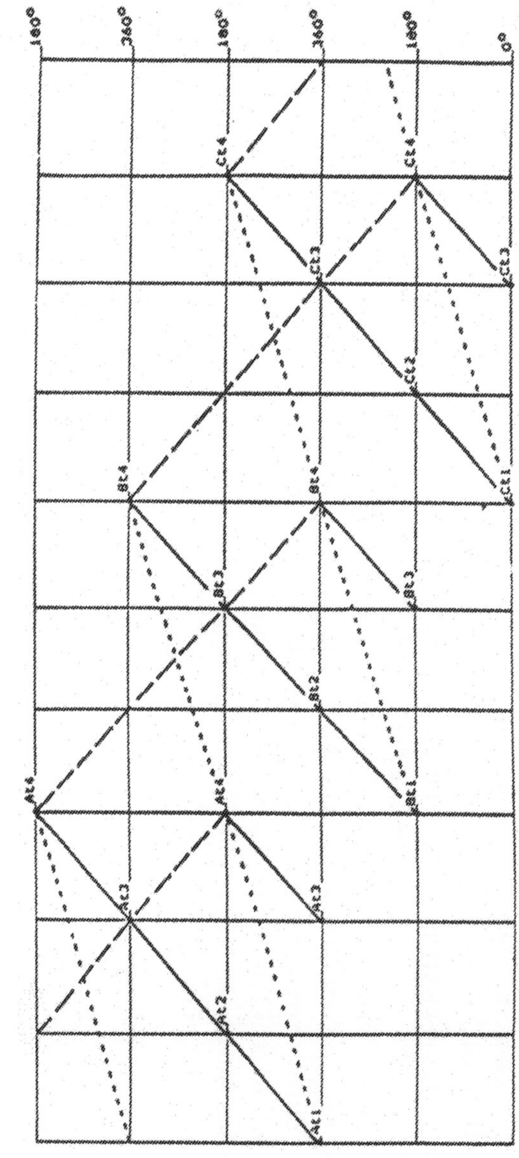

THE CONTRACTION OF THE MAGNETIC DOUBLE HELIX

Drawing 2 When V = $^{1}/_{3}$ C

The value, C is the distance a wave will travel, and the value, V, is the distance that the magnet will move, while the electron is completing one orbit. A wave's length is measured from its origin on an electron to its meeting with a wave front from the other direction. The helical paths of the wave fronts and electrons are shown on a flat paper. Slanted paths on paper become helical when the paper is rolled into a 360 tube. Degrees are shown on the right. The nuclei are not shown. The paths shown are only those made in the period, from time, t1, to time, t5. The paths of the electron orbits, from time, t4, to time, t5, are shown twice. These are identical positions 360 around the tube. $1 - V^2/C^2$ is the formula for the contraction in length in the line of motion. When V/C is $^{1}/_{3}$, V^2/C^2 will be $^{1}/_{9}$. That will be the amount of the overlap and the size of the contraction. The length of a forward moving wave is $^{4}/_{3}$ C or 20 spaces. The length of the reverse wave is $^{2}/_{3}$ C or 6 spaces. C = 9 spaces. At time, t5, electron B will arrive at the junction of waves that left electron A at time, t1, and left electron C at time, t3. The electrons are held firmly between the opposing pulls of the two waves as the intersection circles the tube. The right flowing waves, the forward waves, are shown as dashed lines. The reverse waves are shown as dotted lines. The spiral paths of the electrons are shown as solid lines. There are more details in the text.

Magnetic Double Helix III
DRAWING 2 WHEN V = 1/3 C

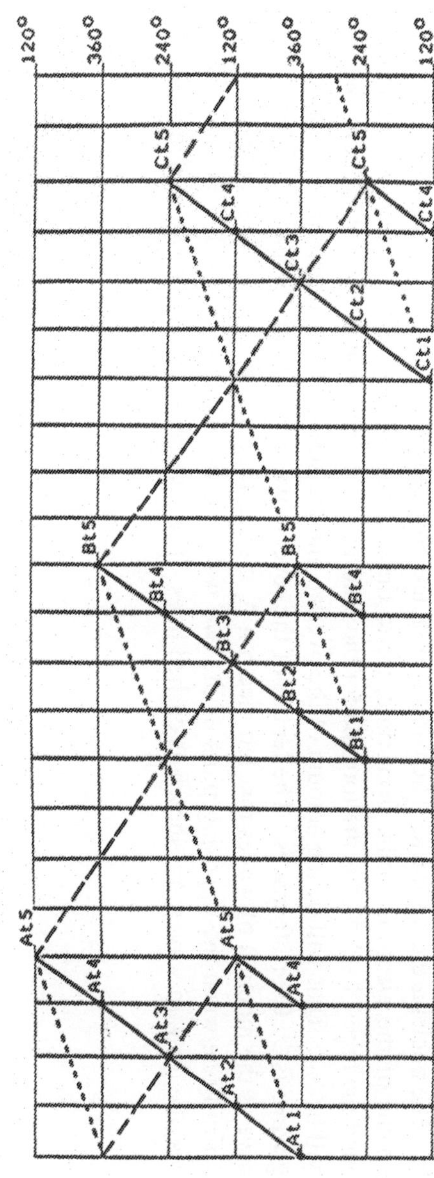

THE CONTRACTION OF THE MAGNETIC DOUBLE HELIX

Drawing 3 When V = ¼ C

The value, C is the distance a wave will travel, and the value, V, is the distance that the magnet will move, while the electron is completing one orbit. A wave's length is measured from its origin on an electron to its meeting with a wave front from the other direction. The helical paths of the wave fronts and electrons are shown on a flat paper. Slanted paths on paper become helical when the paper is rolled into a 360 tube. Degrees are shown on the right. The nuclei are not shown. The paths shown are only those made in the period, from time, t1, to time, t6. The paths of the electron orbits, from time, t5, to time, t6, are shown twice. These are identical positions 360 around the tube. $1 - V^2/C^2$ is the formula for the contraction in length in the line of motion. When V/C is ¼, V^2/C^2 will be $^1/_{16}$. That will be the amount of the overlap and the size of the contraction. The length of a forward moving wave is $^5/_4$ C or 20 spaces. The length of the reverse wave is ¾ C or 12 spaces. At time, t6, electron B will arrive at the junction of waves that left electron A at time, t1, and left electron C at time, t3. The electrons are held firmly between the opposing pulls of the two waves as the intersection circles the tube. The right flowing waves, the forward waves, are shown as dashed lines. The reverse waves are shown as dotted lines.

DRAWING #3 WHEN V = ¼ C

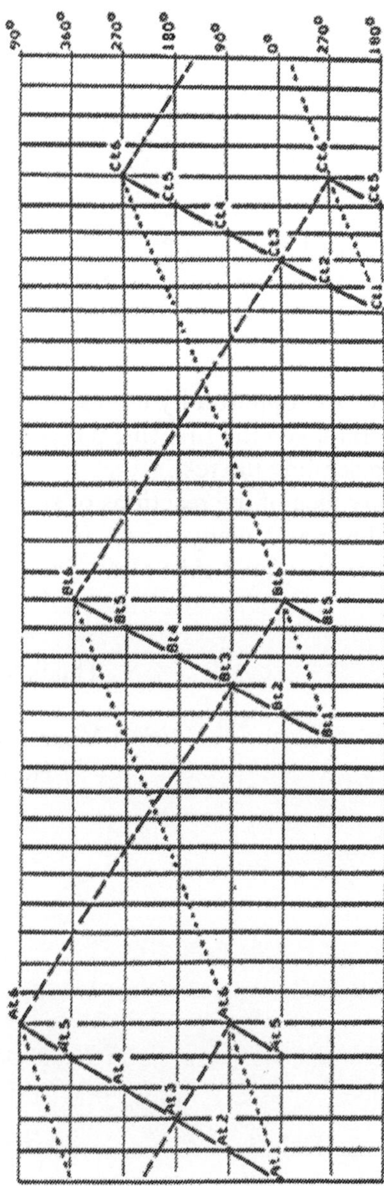

Herman Hagemier
3620 E. 39th St.
Indianapolis, 5, Ind.
Oct. 17, 1962

Dr. Emil J. Konopinski
Indiana University
Dept of Physics

Dear Sir:

I appreciated very much my opportunity to talk to you last Saturday about the theory I am developing. I removed the first five pages because they were much too philosophical and I am sending the rest.

Please don't dismiss my idea of the reactions of spiral waves as a careless hunch. I have gone over this many, many times. If this ether exist, then magnetic repulsion and attraction must exist also. The reactors in the wave could give no other result.

Thank you for any comment
at all.
Yours Truly
Herman Hagemier

Dear Mr. Hagemeier:

I did not return your manuscript at once only because I hate to be brutal toward a lively, although misguided, mind. You would find that any working scientist would consider your suggestions appallingly naive. Surely you cannot believe that this sort of thing is all there is to a genuine effort at clarifying physical theory! Much more sophisticated efforts at making your type of theorizing work were dropped already in the last century, when it became clear that nothing so simple can correspond to the hard facts of nature. You should recognize that you cannot expect a serious hearing until you demonstrate some familiarity with what the problems of theoretical physics really are. Believe me, no one is up to bypassing the thousands of pieces of work by hundreds of perfectly intelligent scientists over more than a century. Their work has at least been far too successful to be ignored, so you ignore it by not even deigning to give it serious study.

Sincerely Yours,

[illegible]

This told me to study 19th Century History

H.H.

www.ingramcontent.com/pod-product-compliance
Ingram Content Group UK Ltd.
Pitfield, Milton Keynes, MK11 3LW, UK
UKHW041933190726
13854UKWH00004B/1562